motortecture
ein jahrhundert architektur und automobilität

motortecture
one century of architecture and automobility

motortecture
ein jahrhundert architektur und automobilität

Wir haben es erfahren und es fast bis zur Selbstaufgabe ertragen: Das Auto ist ein Zerstörer unserer Städte, ein Feind traditioneller Architektur. Kein Wunder, denn diese Städte waren zum Zeitpunkt der Geburt des Automobils auf Maß und Geschwindigkeit eines Ochsenkarrens ausgerichtet und der Dynamik des Autos nicht im Entferntesten gewachsen. Eine Begegnung zwischen archaischer Immobilität und stürmischem Vorwärtsdrang, die Folgen zeigt. Das Auto bedroht und (zer)stört unsere Lebensräume. Wir wehren uns nicht, und – das ist der entscheidende Webfehler – das Auto ist einer unserer vorrangig geliebten Lebensräume. Wenn man es mit der ganzen Wahrheit hält: Es ist das Goldene Kalb der Moderne. Wir haben uns der Automobilität rettungslos ausgeliefert. Und eines ist klar, noch sehr lange werden unsere Gesellschaften in Nord und Süd, Ost und West nicht ohne, sondern nur mit dem Auto existieren wollen und können. Wie die friedliche Koexistenz aussehen kann? Wir wissen es nicht, wir arbeiten daran. Doch zunächst ein Rückblick.

alles auf anfang: automobilität auf dem weg zum mythos und motor der gesellschaft

Die Geschichte mit dem Rad ist jahrtausendealt. Dass es ohne menschliche oder Pferdestärken rollen kann, gerade knapp zwei Jahrhunderte. Einen genauen Startpunkt kann man kaum definieren: „Das Auto wurde nicht erfunden, wie etwa Thomas Edison die Glühbirne oder das Radio erfand. Vielmehr nahm im Automobil langsam eine Idee Gestalt an – die der Automobilität".[1] Klar – am Anfang steht eine Erfindung und der Patent-Motorwagen des Carl Benz. Das war 1886. Aber die Vehikel der Auto-Pionierzeit wirkten skurril;

---

motortecture
one century of architecture and automobility

We have experienced it ourselves, and suffered it almost to the point of self-denial: the car is a destroyer of our cities, an enemy of traditional architecture. And no wonder, for when the car was born those cities were geared to the dimensions and speed of an ox-cart, and not in the remotest prepared for the dynamism of the motor car. This encounter between archaic immobility and an impetuous longing for forward movement has not been without its consequences. The car threatens and disturbs/destroys our habitats. We do not defend ourselves: indeed – and this is where we have a screw loose, because we adore it – the car is one of the habitats we love most. To tell the whole truth: it is the Golden Calf of modernity. We have surrendered ourselves irredeemably to automobility. And one thing is clear: for a very long time to come, our societies north and south, east and west, will neither want nor be able to exist without the car, but only with it. What might peaceful co-existence look like? We do not know, we are working on it. But first, let us have a look at how it all began.

back to square one: automobility's progress to myth and motor of society

The wheel has been around for thousands of years. Only for just under two centuries has it been possible to make it roll without human effort or horsepower. It is barely possible to define a precise point in time when it all began: "the car was not invented in the sense that Thomas Edison invented the light bulb or the radio. Rather, it was in the car that an idea – that of automobility – gradually assumed concrete form."[1] What is clear is that its beginning is marked by an invention and a tangible motor vehicle patented by Carl Benz. That was in 1886. But the vehicles from the pioneering age of the car were weird in

es dauerte lange, bis sie breitere Akzeptanz fanden, zunächst als Sport- oder Luxusartikel, dann allmählich wurden sie auch zum generellen Transportmittel und erst viel später zum Instrument einer Massenmotorisierung. Hier befinden wir uns bereits in den zwanziger Jahren des 20. Jahrhunderts.

1913 hatte Henry Ford mit seiner Fließbandfertigung zunächst für die USA das Automobil in Richtung Massenverkehrsmittel bewegt. Tin Lizzy, sein T-Modell, kostete im Jahr 1924 nur 290 Dollar, nicht viel mehr als das Doppelte eines durchschnittlichen Monatslohns eines Arbeiters damals. An der Wende vom 19. zum 20. Jahrhundert musste für ein Auto etwa noch das Hundertfache der persönlichen Monatseinnahmen aufgebracht werden.

Erst die Bereitstellung für die Massen über alle Klassen- und Schichtengrenzen hinweg hat das Automobil in seine zentrale Funktion zwischen Mythos und Motor der Gesellschaft versetzt. Seine psychologischen Qualitäten („Rollender Uterus" – Peter Sloterdijk) paaren sich mit einem menschlichen Traum der Ubiquität. Für jeden und jederzeit. Natürlich brachte das Auto gemessen an der gesamten Menschheitsgeschichte die Evolution auch nur einen kleinen Schritt voran, aber dafür ging es gewaltig schnell.

Zusätzlich wurde spätestens mit Ford (in Europa mit dem KDF-Wagen, der zum Volkshelden, pardon, Volkswagen wurde) eine bedingte Demokratisierung der Industriegesellschaften losgetreten, die die wirklichen Unterschiede zwischen den Klassen und Schichten einebnete und überdeckte. Denn wer ein Auto hatte, war mobil, konnte am Leben mit gänzlich anderer Qualität teilhaben. Mit dem

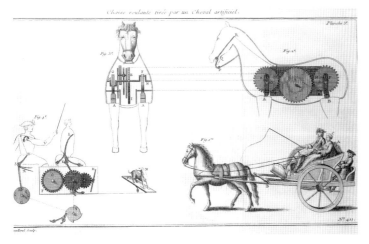

Jede Zeit hatte ihre spezifischen Träume von der Automobilität: Patent für einen Wagen mit automatischem Pferd (zeitgenössische Darstellung, 1735).
Every age has had its specific dreams of automobility. Patent for a cart with an automatic horse (contemporary drawing, 1735).

In der Retrospektive ein skurriler Pionier: der Patent-Motorwagen von Carl Benz, 1886.
Seen from the present day, it was a comical pioneer. The car patented by Carl Benz in 1886.

appearance, and it took a long time until they were widely accepted. First they were sports or luxury articles, then gradually they became a general means of transport, and it was only much later that they became the means of mass motorization. By the time that happened, we were already in the 1920s.

In 1913, Henry Ford and his assembly line production started the process by which the car was to become a means of mass transportation, initially in the USA. Tin Lizzy, his Model T, cost a mere 290 dollars in 1924, not much more than an average worker earned in two months at that time. At the beginning of the same decade, a car had still cost a hundred monthly salaries.

It was only when it became available for the masses across all class barriers that the car found itself in its central function, between myth and motor of society. Its psychological qualities (Peter Sloterdijk calls it a "uterus on wheels") are paired with man's dream of ubiquity. For every one, everywhere. In terms of the history of mankind as a whole, of course, the car only took evolution a short step forward, but that step happened extremely quickly.

Moreover, with the advent of Ford at the latest (or, in the case of Europe, of the "KDF" (strength through joy) car, which became a car for the "volk" (=Volkswagen), if not a popular hero) a certain democratization of industrial societies was set in motion, which hid and papered over the real gaps between classes. Anyone who had a car was mobile, and could share in a completely different quality of life. With a car, you were "in". It was roughly similar to the situation with the internet today. It was only much later (in Germany, for example), following the Economic Miracle,

9

Auto war man „drin". Über den variantenreichen Feinschliff zum differenzierten Statussymbol hat man sich (beispielsweise in Deutschland) erst viel später, nach dem Wirtschaftswunder, Gedanken gemacht.

Es bleibt die Tatsache: Schon zu Beginn des 20. Jahrhunderts hatte sich das Auto bereits in seine gigantische Rolle gefunden, die eine interessante Mixtur aus ersehntem Fortbewegungsmittel und heimlichem Identifikationsobjekt war. Ein sinnstiftendes Vehikel und ein unermüdlicher Wirtschaftsmotor dazu, wie sich inzwischen herausgestellt hat.

das auto ignoriert die architektur, die architektur zunächst auch das auto

Auch den visionärsten unter den Planern und Architekten zu Beginn der modernen Architektur war nicht endgültig klar, mit welch gewaltigem metallischem Prometheus sie es zu tun hatten. Sie ahnten nicht einmal, dass sie möglicherweise von Anfang an nicht die aktiven, sondern reagierende Kräfte waren. Sie wussten nicht, dass durch die Macht des Faktischen alle kühnen Visionen bereits überrollt und obsolet geworden waren. Die Begegnung zwischen Auto und Architektur war keine jungfräuliche mehr, kein Kuss mit folgendem Eintritt ins Paradies, weil auf dem Gebiet des Städtebaus die Macht im Voraus verteilt und festgezurrt worden war, möglicherweise alle gewaltigen Stadttheorien und -utopien schon von vornherein hinfällig waren. Die Helden der Moderne wie Antonio Sant' Elia (entwarf eine „Città Nuova") oder Le Corbusier (eine „Ville Contemporaine") waren keine wirklichen Beweger, sondern

Zwischen Manufaktur und industrieller Produktion:
frühe Autoproduktion bei Mercedes-Benz in Stuttgart (vor 1920).
Between manufactory and industrial production: early car production
at Mercedes-Benz in Stuttgart (before 1920).

that people began to think about the many finishing touches that would set these status symbols apart from each other.

What remains is the fact that, right at the beginning of the 20th century, the car had assumed its monumental role, a bold mixture of longed-for means of locomotion and secret object of identification. What is more, as has meanwhile become clear, it is also a vehicle bestowing meaning and an unflagging motor of economic growth.

the car ignores architecture,
and vice versa (at first)

At the beginning of modern architecture, even the most visionary of planners and architects did not ultimately realize how strong the metal Prometheus was with which they were confronted. They did not even have an inkling that, possibly right from the beginning, they were not acting, but reacting. They did not know that the power of actual facts had already steamrollered all their bold visions and made them obsolete. There was no longer anything virginal about the encounter between car and architecture, no kiss followed by entry into paradise, because in the field of urban planning power had been shared out and finalized in advance, and possibly all the powerful urban theories and utopias were outmoded even before they were born. The heroes of modernism, such as Antonio Sant' Elia (who designed a "città nuova") or Le Corbusier (who designed a "ville contemporaine"), were not real movers, but were barking up the wrong tree right from the start: with their heads in the clouds, they failed to appreciate the actual problems the car presented. We will return to this point later.

von Anfang an auf dem Holzweg, weil sie sich über eigentliche Probleme mit dem Auto hinwegträumten. Darauf wird zurückzukommen sein.

Erst einmal hat Architektur wenig zu tun und im Sinn mit dem Auto, genauer gesagt: Als Bauaufgabe war das Autothema für den künstlerisch und skulpturell begabten Architekten nur wenig attraktiv, denn Wohnungen für Blechkarossen gibt es nicht, allenfalls heißen sie Garagen und sind eine Fortschreibung der Remise, also ein klassisches Neben- oder Nutzgebäude, an dem Repräsentationsgesten nicht erwünscht sind.

Ein Parkhaus besitzt natürlich nicht den Status einer Kathedrale oder eines Schlosses. Obwohl folgende Vorstellung reizvoll ist: Moderne Städte sollten sich nicht nur über ihre Altstadt und die dazugehörigen Kirchtürme definieren, sondern oder vor allem auch über die Architekturqualität oder die Ikonografie der jeweiligen Zentralgarage! Soll heißen: Nicht allein die Türme der Kathedrale sind das Wahrzeichen einer Stadt, sondern auch Art und Kontur der Zentralgarage. Zu einem solchen Denken ist es nie gekommen, was beim heutigen Götzenstatus der Autos erstaunt, denn Parkhäuser könnten so etwas wie die Vitrinen in der guten Stube sein. Wenn Boris Podrecca in Bozen ein riesiges Areal wie die Walther-Passage am Waltherplatz mitten in der Altstadt baut, dann brilliert er mit seiner Baukunst bei Geschäftshäusern und einer funkelnden Galerie. Die Autos aber stehen in einer stickigen Garage unter der Platzsohle. Warum versteckt er die Autos, die gerade in Italien einen absoluten Liebhaberstatus besitzen?

Mit der Automobilität wuchs ein neues Problem heran: kompakten Stauraum in den Städten zu schaffen. Genial einfaches Beispiel: Chicago Parkregal, zwanziger Jahre.
A new problem emerged with automobility: that of finding compact storage space in the city. One brilliantly simple example is this parking shelf in Chicago in the 1920s.

The first thing to say is that architecture has little to do with cars, nor does it want to have much to do with them, or, more precisely: for the artistically and sculpturally gifted architect, the car topic held out little attraction as a building task, because there are no dwellings for old crates, but at best garages, which are a continuation of the shed, i.e. a classical outbuilding or utility building, for which prestigious gestures are not desired.

Of course, a garage does not have the status of a cathedral or palace. Although the following idea is an attractive one: modern cities should define themselves not in terms of their historic centres and appurtenant church towers, but also, if not especially, in terms of the architectural quality or iconography of their respective main garage. In other words: it is not just cathedral towers that are the distinguishing feature of a city, but also the type and contours of its main garage.

Astonishingly, given the golden calf status cars enjoy today, nobody has ever had this idea, and yet garages could be like glass showcases in the front parlour. If someone like Boris Podrecca builds a huge complex like the Walther-Passage mall on Waltherplatz, in the historic centre of Bolzano, he displays the brilliance of his architectural art in business premises and a sparkling gallery. Cars are stuck in a stuffy garage below the base of the square. Why does he hide the cars? And especially in Italy, where they enjoy absolute cult status?

But there are exceptions – at the Stedelijk Museum in Amsterdam or in Lyon, where garages are no longer black holes, but are lovingly designed. This begins with a careful selection of refined flooring materials and ends with the integration of remains of the old town walls and of buil-

Ausnahmen gibt es allerdings: In Amsterdam am Stedelijk-Museum oder in Lyon, wo Parkhäuser keine dunklen Löcher mehr sind, sondern liebevoll gestaltet werden. Das beginnt mit der sorgfältigen Auswahl gepflegter Bodenbeläge und endet mit der Integration alter Stadtmauerreste und Hausfundamente – als Teil der Stadt unter der Stadt. Dafür zahlt man als Besucher gern ein bisschen mehr.

den schneid abgekauft:
das dampfermotiv der moderne

Es gibt keine ausgeprägte, keine gesicherte Typologie von Bauten, die zum Thema Auto gehören. In jedem Fall nicht so ausgeprägt wie anderswo, wo sich eine zusammenhängende Baugeschichte ergibt wie bei Bankpalästen oder Bahnhöfen. Busbahnhöfe wie auch Parkgaragen oder Tankstellen und Feuerwehrstationen sind über den Status eines Nutzbaus nicht hinausgekommen, die Stararchitekten haben sich diesen Aufgaben nicht wirklich gewidmet, ausgenommen vielleicht der Amerikaner Frank Lloyd Wright mit einem Tankstellenentwurf. Oder Robert Venturi mit seinen Feuerwehrstationen, der Fire Station No. 4 in Columbus, Indiana und der Dixwell Feuerwache in New Haven, Connecticut. Aber das war erst in den sechziger und siebziger Jahren. Parkhäuser sind uns am liebsten, wenn sie unter der Erde sind (s.o.). Dabei besaßen die ersten Auto-Regale, sei es in Paris oder in Chicago, durchaus einen hohen Reiz, vergleichbar mit der ausgefeilten Ingenieurskunst einer Schiffshebewerkskonstruktion.

Es gibt also kaum eine gesicherte Autoarchitektur-Typologie, allerdings wird andersherum ein Schuh daraus. Die Architekten schauten

Sie blieben überraschenderweise Stiefkinder der modernen Architektur: Bauten fürs Auto, hier eine Tankstelle der DAPG (Deutsch-Amerikanische Petroleum-Gesellschaft, ca. 1930).
Surprisingly, buildings for cars were to remain neglected by modern architecture: the picture shows a DAPG (Deutsch-Amerikanische Petroleum-Gesellschaft) filling station, about 1930.

ding foundations – as part of the city under the city. Clearly, users have no objection to paying a little more for parking there.

There is no distinctive, definite typology of buildings associated with the car theme. Or, at all events, it is not as distinctive as in other areas with their own cohesive architectural history, such as bank headquarters or stations. Bus stations and garages, filling stations and fire stations have never progressed beyond the status of a utility building, and star architects have never truly devoted themselves to such tasks, with the possible exception of the American Frank Lloyd Wright and his design for a filling station. Or Robert Venturi, with his two fire stations, Fire Station No. 4 in Columbus, Indiana and Dixwell Fire Station in New Haven, Connecticut. But that was not until the 1960s or the 1970s. We like garages best if they are underground (see above). And yet, the first car silos, whether in Paris or in Chicago, were without doubt extremely attractive, comparable with the sophisticated engineering art of a barge lift.

stealing their thunder:
modernism's steamer motif

There is, then, hardly any definite car architecture typology, but then again exactly the opposite is the case. Architects took a close look and, from the semantic world of steam locomotives, steamships and cars, they stole what they could get their hands on, using it for houses or villas, for banks and other buildings. Less was done for a graceful garage. Modernism got extremely excited about the "steamship motive",[2] becoming addicted to the filigree forms of ships rails and the powerful geometry of a ship's hull. Looked at from this angle, there is a great deal

Moderne Bauten, die vom Abschied des Alten künden
mit neuen Autos, die uralt ausschauen: Weißenhofsiedlung
Stuttgart mit einem Mercedes-Benz (Architektur:
Le Corbusier).
Modern buildings proclaiming a farewell to tradition, with
new cars that look ancient. Weissenhof apartments in Stuttgart
with a Mercedes-Benz (architecture: Le Corbusier).

genau hin und kopierten aus der semantischen Welt der Dampf-rösser, Dampfer und Autos, was sie bekommen konnten und be-nutzten es für Wohnhäuser oder Villen, für Banken oder Sonstiges. Weniger für eine anmutige Garage. Man berauschte sich am „Dampfermotiv",[2] verfiel den filigranen Formen von Relings und der Kraft der Geometrie eines Schiffsbaukörpers. Wenn man es so be-trachtet, existiert sehr wohl eine fruchtbare Verbindung zwischen Technik (darunter das Auto als Teil davon) und Baukunst.

Wer die wunderschönen Fotos von Frank Yerbury kennt, seine Foto-grafien von Villen des Architekten Le Corbusier in Verbindung mit stattlichen Karossen oder dem Citroën-Showroom in Paris, dem wird auffallen müssen, dass am Ende der zwanziger Jahre die moderne Architektur allgemein einen technisch-dynamischen Ausdruck gefunden hat. Dass sie mutig mit dem erweiterten Repertoire von Form und Material umging, dass bereits Abstraktion und Entmateria-lisierung Einzug ins architektonische Denken gehalten hatten. Ob Dampfer- oder Automotivik – diese Bauten künden vom Abschied des Alten, von Aufbruch und Zukunft, denn sie sehen so ganz anders aus als ihre Vorgänger.

Die meisten Autos wirkten dagegen immer noch so, als seien sie deformierte Kutschen und stammten aus, wenn auch hochqualifi-zierten, Manufakturen. Eine grobe Täuschung und Verkehrung der Verhältnisse: Was futuristisch aussieht, ist es gar nicht so sehr, in jedem Fall ist es immobil und statisch, erdverbunden usw. Was mobil und dynamisch ist, erinnert noch viel zu sehr, wenn nicht an Plüsch und Plunder, dann an den Formenkanon einer untergegangenen Welt.

of artistic cross-fertilization between technology (of which the car is a part) and architecture.

Whoever knows the beautiful photographs by the photographer Frank Yerbury and his shots of villas by the architect Le Corbusier with splendid limousines in the foreground of the Citroën showroom in Paris must necessarily by struck by the fact that, at the end of the 1920s, modern architecture in general expressed itself in technological and dynamic terms. He will also notice that it made bold use of its extended repertoire of form and materials, and that abstraction and dematerialization had already found their way into the architectural mindset. Whether they use steamship or car motifs, these buildings manifest a farewell to the tradi-tional, a new start and a future world, so different are they from their predecessors.

By contrast, most of the cars of the time still looked like deformed carriages from artisan's workshops, albeit highly qualified ones. A gross delusion, which turns reality on its head: What looks futuristic is not really that futuristic at all: at all events it is immobile and static, earth-bound, and so on. That which is mobile and dynamic is still far too reminiscent, if not of pomp and circumstance, then of the formal canon of a defunct world.

a beautiful example:
lingotto. car-cum-ship in architectural form

There are nevertheless a handful of buildings that can be regarded as key structures of modern car-architecture. For example, Fiat's Lingotto plant in Turin (built between 1924 and 1926) by the engineer Giacomo

ein schönes beispiel:
lingotto, das gebaute auto-schiff

Es gibt dennoch eine Hand voll Bauten, die als Schlüsselobjekte der modernen Autoarchitektur gelten können. Zum Beispiel die Fiat-Fabrik Lingotto in Turin (gebaut zwischen 1924 und 1926) des Ingenieurs Giacomo Mattè-Trucco und deren Metamorphose zum Ende des 20. Jahrhunderts. Die Autofabrik mit direkt zugeordneter Teststrecke auf dem Dach ist ein städtebaulicher Idealfall: Das Auto in der Erprobung auf dem Dach verbraucht natürlich keinen zusätzlichen Platz.

Lingotto ist eines der frühen Beispiele für eine modulare, das heißt baukastenmäßige Konstruktion aus Eisenbeton. Nur drei Grundelemente, nämlich Pfeiler, Balken und Decke, werden zur Komposition benötigt. Weil hier schon in den zwanziger Jahren die größte komplette europäische Autoherstellung ihre Heimat fand, kann man durchaus von einer ersten integrierten Auto-Stadt sprechen. Besonders die Teststrecke auf dem Dach und ihre expressiv symbolische Führung der Betonrampen für die Auffahrt der Wagen bringen Lingotto zum Sprechen, es wird buchstäblich zur semantischen Auto-Tektur. Die gewaltigen Fortschritte bei der serienmäßigen Fertigung von Eisenbeton ermöglichten diese eben sehr große Geburtsstätte der neuen Blechkarossen. Lingotto war und ist deswegen „unkaputtbar" und wurde trotz heftiger alliierter Angriffe nicht einmal im Krieg wirklich zerstört. Es war allerdings später für seine ursprünglichen Aufgaben unbrauchbar geworden und wurde in den achtziger Jahren in Rente geschickt, weil modernste Autoproduktion anderswo kostengünstiger und komprimierter darstellbar ist.

Städtebaulicher Idealfall: Fiat-Fabrik Lingotto mit Teststrecke auf dem Dach (1924–26).
Perfect urban planning: Fiat's Lingotto plant with race track on its roof (1924–26).

Lingotto nach der Konversion: Multikomplex für Messe, Museum und vieles mehr (1983, Architekt: Renzo Piano).
Lingotto plant following conversion. A multiple-use complex for trade fairs, museums and much more (1983. Architect: Renzo Piano).

Mattè-Trucco, and its metamorphosis at the end of the 20th century. The car plant with a race-track on its roof is an urban planning ideal: no additional space is used up if the roof is used to put the car through its paces.

Lingotto is one of the early examples of modular, i.e. kit-like construction, using reinforced concrete. Only three basic elements – piers, beams and slabs – are needed for this composition. Because Europe's largest complete car plant was given a home here are early as the 1920s, we can by all means speak of it as the first integrated motor town. In particular, the race track on the rook and the expressive symbolic lines of the concrete ramps up which the cars drive give Lingotto its resonance, its truly autotectural semantics. The huge progress made in the series production of reinforced concrete makes possible what is, when all is said and done, a huge birthplace of new limousines. This is why Lingotto was and is indestructible, and was not even properly destroyed in the war, despite heavy shelling by the Allies. However, it had become unusable for its original task and was put out to grass in the 1980s, as ultra-modern car production can be done elsewhere less expensively and in less space.

Star architect Renzo Piano (who meanwhile has a further long-standing business relationship with Mercedes-Benz and DaimlerChrysler) hit the nail on the head: "Because of its size and economic role, Lingotto had always been an urban district of its own".[3] It could neither be demolished nor trivialized. The idea was to preserve this autotecture as a place of work and a cultural-historical site. Piano's response was a mixture of uses from culture and business, with trade fair, museum and concert hall. In the meantime, the old car plant has itself provided

Vorbildliche Türme für Menschen und Autos: Marina City in Chicago (1959–64, Architekt: Bertrand Goldberg).
Exemplary towers for people and cars: Marina City in Chicago (1959–64. Architect: Bertrand Goldberg).

Stararchitekt Renzo Piano (der inzwischen mit Mercedes-Benz und DaimlerChrysler einen weiteren Autobauer als Dauerkunden hat) erkannte ganz richtig, dass „Lingotto sowohl seinen Dimensionen als auch seiner ökonomischen Rolle nach ein Stadtteil"[3] war. Weder abreißen noch verniedlichen war möglich. Diese AutoTektur sollte weiterhin Arbeitsplatz und kulturhistorischer Faktor bleiben. Piano hat mit einer Nutzungsmischung aus Kultur und Wirtschaft, aus Messe, Museum und Konzertsaal geantwortet. Inzwischen ist die alte Autofabrik zum Katalysator der weiteren Entwicklungen im ganzen Stadtteilgebiet geworden. Die Zubauten Pianos auf dem Dach mit gläsernen Kuppeln lassen den Gesamtkomplex Lingotto wie einen Flugzeugträger wirken; schon zu Lingottos Gründungszeiten hatten zeitgenössische Chronisten die Fabrik als Dampfer an der Reede bezeichnet. Die Bilder der frühen Industriezeit vermischen sich jetzt mit neuen.

**marina city in chicago:
twin towers fürs auto**

Zwischen 1959 und 1964, etwa zehn Jahre früher als das New Yorker World Trade Center, wurden in Chicago Zwillingstürme fertiggestellt, die ihren Architekten über Nacht weltberühmt machten, so berühmt, dass er kaum noch etwas anderes zu bauen brauchte. Marina City war revolutionär, und niemand denkt heute noch darüber nach, dass viele modische Adaptionen von heute eigentlich auf sie zurückgehen: Die Türme sind rund, nicht rechteckig, stehen allerdings in der Bannmeile des Königs des rechten Winkels, direkt neben dem Seagram Building von Mies van der Rohe. Was wie „Gotteslästerung" aussieht, ist nur dem Auto geschuldet. Und das

the catalyst for further developments throughout the district. Piano's extra structures on the roof, with their glass domes, give Lingotto as a whole the appearance of an aircraft-carrier. Indeed, even when it was built, contemporary observers had likened the plant to a steamship lying in the roads. The images from the early years of industrialization now mix with new ones.

marina city in chicago:
twin towers for the car

Between 1959 and 1964, ten years before New York's World Trade Center, twin towers were completed in Chicago. They made their architect world-famous overnight – so famous that he scarcely ever needed to build anything else. Marina City was revolutionary, yet today nobody remembers that this is really the point from which many of today's fashionable adaptations started: its towers were round, not rectangular, even though they stood in the inviolable precincts of the King of the Right Angle, directly next to Mies van der Rohe's Seagram Building. What appears to be sacrilege is due solely to the car. This was how it came about: Bertrand Goldberg wanted to have a city within the city: apartments for singles and dinks, combined with other uses for the Loop, as Chicago's down town area is known. As the architect Bertrand Goldberg could not stack cars in front of these residential towers, he collected them in the basement, on the lower storeys, where they can easily be seen, as the Cadillacs and Buicks were on open view behind curved precast concrete sections.

Perhaps intuitively, perhaps also as a result of his experience, he combined uses for leisure, for boats and cars with living and working

kam so: Bertrand Goldberg wollte eine Stadt in der Stadt: Wohnungen für „Singles" und „Dinks" (Double Income No Kids) kombiniert mit anderen Nutzungen für den Loop, der Downtown von Chicago. Da der Architekt Bertrand Goldberg die Autos mitten in der Stadt nicht vor den Wohntürmen der Menschen aufstapeln konnte, versammelte er sie im Keller, in den unteren Geschossen, gut sichtbar für jedermann, da die Cadillacs und Buicks offen hinter schwungvollen Betonfertigteilen standen.

Vielleicht intuitiv, vielleicht auch aufgrund seiner Erfahrungen kombinierte er Nutzungen für Freizeit, für Boote und Autos mit Wohnen und Gewerbe auf kreisförmigem Grundriss. Er hatte herausgefunden, dass sich nicht nur Wohnungen auf diese Weise optimal erschließen lassen, sondern er außerdem Stützen und Material einsparte und natürlich auch kreisrunde Spindeln und Rampen für die Fahrradien der Autos ideal waren. Ganz nebenbei, wahrscheinlich aber sehr bewusst, hatte er die Trennung der Funktionen in der modernen Stadt wieder aufgehoben, hatte autogerecht und gleichzeitig urban gedacht und eine prägnante Großform entwickelt. Bertrand Goldbergs Bau ist ein Glücksfall für die moderne Architektur, insbesondere im Rahmen dieses Themas einer Auseinandersetzung mit Automobil, Architektur und Stadt. Er hat Autos und Menschen in einem Gebäude versöhnt, dank seiner dynamischen Architektur ist ihm ein Klassiker gelungen. Diesem großartigen Original sind leider nur wenige gelungene Kopien gefolgt.

Warum er als Architekt nicht nachhaltig erfolgreich war, hat vermutlich einen einfachen Grund: Bertrand Goldberg war schlichtweg seiner Zeit und deren Architektur voraus und möglicherweise am

on a circular ground plan. He had found out that this was not only an excellent way of connecting apartments to utilities and services, but also that he saved on supports and materials, and of course that circular spirals and ramps were ideal for the cars' turning radius. Quite incidentally, but probably very deliberately, he had abandoned the separation of functions in the modern city, using concepts that were car-friendly as well as urban, and had developed a succinct monumental form.

Bertrand Goldberg also produced a stroke of modern architectural luck, particularly in our context of the interplay between car, architecture and city. He reconciled cars and people in one building, and thanks to that building's dynamic architecture he succeeded in creating a classic. Unfortunately, only a few buildings have managed to emulate this shining example.

It is probably simple to explain why his success as an architect did not last: Bertrand Goldberg was simply ahead of his time and its architecture, as well as probably being in the wrong place. In the 1970s, when asked by the German architectural researcher Heinrich Klotz why, after having built Marina City, he did not have a firm place in the textbooks of architectural history, unlike Philip Johnson or Louis Kahn, he replied: "Books like those are usually only written about our trendy East Coast architects!"[4]

landmarks on the horizon, glass showcases
and other construction sites

Marina City's towers are an early example of a mixed-use building in which cars play an important role. Usually, however, they disappear into

Prototyp für den Auto-Showroom: Citroën in Paris
(1929, Architekten: Laprade und Bazin).
Prototype of the car showroom: Citroën in Paris
(1929. Architects: Laprade and Bazin).

falschen Ort! Sagte er doch in den siebziger Jahren dem deutschen
Architekturforscher Heinrich Klotz auf die Frage, warum er an-
gesichts von Marina City in den Geschichtsbüchern der Architektur
keinen festen Platz einnähme wie etwa Philip Johnson oder Louis
Kahn: „Meist werden solche Bücher doch nur über unsere Mode-
architekten an der Ostküste geschrieben!"[4]

zielpunkte am horizont, gläserne vitrinen und andere baustellen

Die Marina City ist ein frühes Beispiel eines „Mixed-Use-Building",
in dem Autos eine wichtige Rolle spielen. Üblicherweise verschwinden sie meist in den Keller. Oder werden, wo wie in Amerika viel
Platz ist, einfach auf landfressenden Großparkplätzen abgestellt.
Hier formulieren sie keinen eigenen Ausdruck für AutoTektur.

Eine der wenigen ganz eigenen und typischen Entwicklungen
ist vielleicht an der Nahtstelle zwischen Aufbewahrung und Präsentation das, was heute Auto-Showroom genannt wird (vgl. *corporate
architecture*). Eines der ersten Beispiele dafür war der ingeniöse
Showroom von Citroën in Paris, bei dem sich Betonkonstruktion und
Limousinen optisch auf kunstvolle Weise verwoben – zum Auto-Haus
im buchstäblichen Sinne.

Weil aber ein Verkaufsgeschäft für Autos zwangsläufig viel mehr
Platz benötigt als ein Juwelier, wandelten sich Autohäuser nach
dem Zweiten Weltkrieg vom experimentierfreudigen und exklusi-
ven Verkaufsbauwerk wie an der Düsseldorfer Königsallee oder
am Hamburger Ballindamm zum gesichtslosen Gewerbebau an der

the basement or, if there is a lot of space as in America, they are simply
parked on huge, land-devouring car parks. In this way, they cannot formulate any autotectural expression of their own.

One of the few quite specific and typical developments is perhaps the
one to be found in the interstice between garaging and presentation –
in what is today called the car showroom (see *corporate architecture*).
One of the first examples of this was the ingenious Citroën showroom in
Paris, in which concrete structures and limousines were optically inter-
woven in elaborate fashion to form what was literally a car house.

However, because a car-selling business necessarily takes up more room
than a jeweller's, car showrooms in the post-war period turned away
from experimental and exclusive sales premises, like those found in
Dusseldorf's Königsallee or Hamburg's Ballindamm, becoming faceless
commercial buildings on the periphery of cities instead. It is only now
that there are signs of them returning (see *corporate architecture*, Saab
showroom).

A further type of building relevant to our topic is the bridge-top restaurant crossing the motorway. This unites the bridge theme with a catering
function. Ponte Vecchio in Florence was the first example of how to
combine the traffic-linking function with a market place. If restaurants
are built on bridges across motorways, then only one restaurant is
needed instead of one on each side, and the restaurant itself forms the
link to the other side. If its architects design it with sufficient care,
the restaurant can be a landmark for the drivers passing by, a destination,
a stopover on their long journey and, in the rear-view mirror, a sign of
how far they have already travelled. The more impressive and easier to

Orientierungspunkt für Autobahnfahrer:
Brückenrestaurant Dammer Berge.
Landmark for drivers: bridge-top restaurant
Dammer Berge.

Peripherie der Städte. Erst jetzt ist wieder eine Rückkehr zu vermelden (vgl. Saab-Showroom, *corporate architecture*).

Eine weitere interessante Spezies dürften Brückenrestaurants über Autobahnen sein. Sie vereinen den Bautypus Brücke mit der Funktion einer Gastronomie und dergleichen. Schon der Ponte Vecchio in Florenz hatte ja vorgemacht, wie man die verbindende Verkehrsfunktion mit einem Marktplatz kombinieren kann. Bei den Autobahn-Brückenrestaurants braucht man also nur eins statt zweier Restaurants auf jeweils einer Seite zu bauen, die Verbindung zur jeweiligen anderen Seite bildet das Restaurant selbst. Das Restaurant wird, wenn die Architekten es entsprechend gestalten, zum Merk- und Orientierungspunkt des autofahrenden Passanten, als Ziel, als Zwischenstation seiner langen Reise und im Rückspiegel zum Kontrollinstrument seines bereits zurückgelegten Weges. Je charakteristischer und imposanter die Kontur, desto deutlicher die Erinnerung. Brückenrestaurants gibt es in Frankreich und Italien, in Deutschland leider nur eine Hand voll.

gedanken einer lichtgestalt:
„die city der geschwindigkeit ist eine erfolgs-city"

Ob es nun parkt oder fährt, das Automobil trägt in jedem Fall eine neue Dimension in die Architektur, eben in die Stadt. Die charmanten, aber rettungslos in sich verwachsenen mittelalterlich geprägten Altstädte sind nicht geeignet, auch nicht die gründerzeitlichen Quartiere mit ihren vielen Höfen, die mit Autos oder Lastwagen nur schlecht erreichbar waren. Die Architekten der Moderne wollten das ändern und die neue, die autogerechte Stadt bauen. Das ist an

remember its features are, the more vivid the memory will be. Bridge-top restaurants are to be found in France and Italy, but unfortunately there are only a handful of them in Germany.

thoughts of a leading light:
"a city made for speed is made for success"

Whether parked or driving, at all events the car brings a new dimension to architecture, or rather to the city. The charming yet irredeemably tangled mediaeval centres of cities were not ready for them, yet neither were the proto-industrial quarters with their many courtyards, which could only be accessed with difficulty by cars or lorries. The architects of modernism wanted to change this and build a new, car-friendly city. This can best be illustrated here by a very brief look at the architectural luminary Le Corbusier (1887–1965). Born as Charles-Edouard Jeanneret in western Switzerland, his work as an artist-architect places him without doubt (and not only among the community of his admirers) head and shoulders above all the master-builders of the 20th century. However, he was criticized and indeed almost hated for his role as a visionary and thus radical urbanist, because both his theoretical concepts such as the "ville contemporaine" (1922) or its variant for Paris, the "plan voisin" (1925), or his colossal later plan for a "ville radieuse" (1930) as well as his concrete proposals for Algiers or Chandigarh had, in their grandezza and geniality, in the end been planned without considering the human factor — or at least the type of person as he existed in the 20th century, together with his peculiarities.

The reason may simply be that Le Corbusier worked according to the logic of a mechanical engineer and his ideas of optimization, and so was

dieser Stelle und in aller Kürze wohl am besten an und mit der architektonischen Lichtgestalt Le Corbusier (1887–1965) zu verdeutlichen. Der gebürtige Westschweizer, der mit bürgerlichem Namen Charles-Edouard Jeanneret hieß, ist mit seiner Leistung als Künstler-Architekt nicht nur bei seiner Fangemeinde die unbestrittene Nummer eins der Baumeister des 20. Jahrhunderts. Kritik und beinahe sogar Hass hat er sich allerdings in der Rolle als visionärer und deswegen radikaler Urbanist eingehandelt. Sowohl seine theoretischen Konzepte wie die „Ville Contemporaine" (1922), deren Pariser Variante „Plan Voisin" (1925) oder später das Mega-Schema „Ville Radieuse" (1930) als auch seine konkreten Vorschläge für Algier oder Chandigarh waren letztendlich mit Grandezza und Genialität am Menschen vorbeigedacht worden – zumindest am existierenden Menschentyp des 20. Jahrhunderts und seinen Eigenarten.

Vielleicht liegt es einfach daran, dass Le Corbusier mit der Logik eines Maschinenbauers und dessen Optimierungsdenken vorging, sich also lieber Wohnmaschinen als Gartenstädten widmete und zu radikal, zu inhuman und letztendlich zu ignorant den prometheischen Kräften des Autos gegenüber war. Dass er aber dem Mythos Auto verfallen war, also unter den Maschinen die Kraftwagen eine wichtige Rolle in seinem Planungs-Koordinatensystem spielten, lässt sich unter vielen Aspekten allein an der von ihm und seinem Vetter Pierre Jeanneret entworfenen „Voiture minimum" erkennen, einem eiförmigen, ökonomisch optimierten Kleinwagen, der fast so etwas wie der Urahn all jener Volkswagen und Citroëns der fünfziger und sechziger Jahre sein könnte. Parallel dazu schuf Le Corbusier auch eine „Maison minimum". In seinem persönlichen „Cabanon" in Roquebrune Cap Martin am Mittelmeer hatte er in

Eiförmig, ökonomisch, optimiert, Urahn von Käfer und 2CV: „Voiture minimum" (1936, Le Corbusier).
Egg-shaped, economical, optimized, the progenitor of the Beetle and the 2CV: "voiture minimum" (1936. Le Corbusier).

more committed to "machines for living" than to garden cities and was, in his dealings with the Promethean forces of the car, too radical, too inhuman and, in the end, too ignorant. To see that he was in thrall to the car as myth, and that the car of all machines played an important role in his system of planning and coordinates, it is in many respects sufficient solely to look at the "voiture minimum" designed by him and his cousin Pierre Jeanneret. This egg-shaped, economically optimized small car could almost be the forefather of all the Volkswagens and Citroëns built in the 1950s and 1960s. Parallel to this, Le Corbusier also proposed a "maison minimum". In his own personal "cabanon" at Roquebrune Cap Martin on the Mediterranean littoral, he had, in a kind of wooden caravan, optimized for himself and the last years of his life all the processes of living, just like in a car.

As architect and urban planner, Le Corbusier saw the car as a pacemaker and inspiration. In fact, it is impossible to give Le Corbusier's work in this context the appreciation it deserves, for any description will more or less be taken out of its wider context. What can be said is that the car is processed in many different ways in his architecture, and that he is not simply content with superficially processing the aesthetic topics of steamer, car or aeroplane, but also incorporates their processes of rationalization in the design and production process.

Without doubt, one such example is his "Maison Dom-Ino" of 1914, whose simple system of supports and add-on form not only alludes to the flexible way a domino can be placed during a game but also, like Tin Lizzy, is inexpensive, and can be built in large quantities, mainly by semi-skilled workers, and finally become the series-produced, elementary backbone of his new ideal city. Didactic and motivating, yet always

einer Art hölzernem Wohnwagen für seine letzten Lebensjahre alle Wohn- und Lebensabläufe extrem reduziert und ähnlich wie in einem Auto gestaltet.

Der Architekt und Stadtplaner Le Corbusier hat das Auto intensiv als Takt- und Ideengeber genutzt. Eigentlich ist es unmöglich, diesen Aspekt im Werk eines Le Corbusier entsprechend zu würdigen, weil alles, was beschrieben wurde, mehr oder weniger aus einem größeren Zusammenhang gerissen ist. Festzuhalten bleibt die Vielschichtigkeit der Verarbeitung des Autos in seiner Architektur, die nicht stecken bleibt in vordergründiger Ingenieursästhetik zum Thema Dampfer, Auto oder Flugzeug, sondern auch deren Rationalisierungsprozesse im Entwurf und der Fertigung übernimmt.

Ein solches Beispiel ist sicher die „Maison Dom-Ino" von 1914. Ihr einfaches Stützsystem und dessen addierbare Form erinnern nicht nur an die Flexibilität eines Domino-Steins. Es kann auch preiswert wie Tin Lizzy in hohen Stückzahlen größtenteils von angelernten Arbeitern gebaut werden und schließlich seriell und elementar zum Rückgrat des, seines, neuen Stadtideals werden. In seiner belehrenden wie motivierenden und auch immer schwärmenden Art sah Le Corbusier in seinem neuen Haustypus ein „Werkzeug und ein Typenhaus, das gesund ist (auch sittlich gesund) und ebenso schön wie die Werkzeuge der Arbeit, die unser Dasein begleiten".[5]

Vieles dieser architektonischen Grundarbeit nimmt Le Corbusier mit hinein in seine Villen und seine „Unités d'habitation", wie er seine gebauten Wohnmaschinen in Marseille oder Berlin nennt. Vieles, wie die Aufständerung seiner Häuser durch „pilotis", die

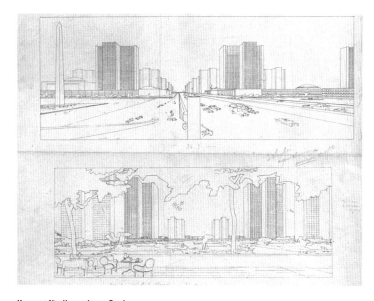

Konzept für die moderne Stadt:
Ville contemporaine (1922, Le Corbusier).
Concept for the modern city:
Ville contemporaine (1922. Le Corbusier).

enthusiastic, Le Corbusier regards this new type of building as a "tool and a model building that is healthy (also morally healthy) and just as beautiful as the tools of work that accompany our existence".[5]

Le Corbusier incorporates a lot of this elementary architectural work in his villas and "unités d'habitation", as he calls the machines for living he builds in Marseille or Berlin. Much in these buildings owes a lot to the car, such as the way he places his buildings on piles ("pilotis"), which make the ground floor into an airy utility floor as it were. Here, cars can park under the buildings or delivery vehicles can bring supplies to apartments.

And there is another typical dominant design feature of Le Corbusier's that comes from the world of the car. In his buildings, Le Corbusier does not work as much with steps as with ramps, which allow people to grasp the spatial context of his buildings in an almost floating way. Here, the pedestrian is confronted with the perception of the coasting car driver.

The avant garde of modern architecture, and Le Corbusier with them, are not so much concerned with buildings and construction as with vision and theory: their focus of concern is not the building but the city – the "car-scaled" city, which will never exist in that form.

It may well be that Le Corbusier's thoughts are not so much directed towards the bright, green city, which everyone wanted to replace the traditional, unhealthy 19th century city, as economically and entrepreneurially motivated: "A city made for speed is made for success."[6] Le Corbusier's plans for a "ville radieuse" are conceived as the first major

sozusagen das Erdgeschoss zum nutzbaren Luftgeschoss werden
lassen, ist direkt dem Auto geschuldet. Hier also können unter den
Häusern Autos parken oder Lieferfahrzeuge die Wohnungen versorgen.

Auch ein anderes typisches raumgreifendes und gestalterisches
Element von Le Corbusier stammt aus der Welt der Kraftwagen.
Le Corbusier arbeitet in seinen Häusern weniger mit Treppen als
mit Rampen, die es den Menschen erlauben, fast schwebend die
Raumzusammenhänge seiner Häuser zu begreifen. Hier wird der
Fußgänger mit der Perspektive eines gleitenden Autofahrers
konfrontiert.

Die Avantgarde der modernen Architektur, und damit eben auch
jener Le Corbusier, bleibt nicht so sehr im Objektbereich, also im
Hausbau gefangen, sondern in der Vision und Theorie: Ihr Fokus ist
nicht das Haus, sondern die Stadt – die „autogerechte" Stadt, die
es dann so nie geben wird.

Le Corbusier denkt dabei möglicherweise nicht so sehr nur an
die lichte und grüne Stadt, die ja nach dem Willen aller die über-
kommene, krankmachende Stadt des 19. Jahrhunderts ablösen
soll, sondern auch wirtschaftlich und mit dem Sinn eines Unter-
nehmers: „Eine Stadt, die für Schnelligkeit gemacht ist, eine Stadt,
die für Erfolg gemacht ist!"[6] Le Corbusiers Pläne für eine „Ville
Radieuse" sind als erste große Korrektur und Antwort auf die
stadtzerstörerische Kraft des Autos gemünzt, sie wollen das Auto
nutzen zum industrialisierten Aufbau. Viele strukturelle Muster,
Aufstellungen und Grundrisse folgen der Logik der optimalen

corrective and response to the car's anti-urban forces, aiming to use the
car for industrialized construction, with many structural patterns, instal-
lations and plans being based on the requirements for optimum crane
positions during construction – something that was even copied later in
actually existing socialism and its buildings made of prefabricated slabs.

car, road, landscape:
the dawn of a new age

Because the intellectual and creative forces of the interwar years are
already – wondrously, even without internet and computers – networ-
ked, the ideas of the avant garde become the textbook knowledge, as it
were, of every planner and engineer in the industrialized world. At the
CIAM (Congrès Internationaux d'Architecture Moderne) congress in
Athens in 1933, a Charter of Athens is passed, in which Le Corbusier
is heavily involved and which is intended to sanction "car-scaled urban
planning". It was intended to be valid for at least the next 100 years.
Like a law, the Charter lays down the cornerstones of development. On
the topic of traffic, for example, it states: "Traffic routes must be classified according to their character and built to match vehicles and their
speed". Or again: "The pedestrian must use different streets than the
car".[7] Separation by function, into a residential, commercial, working and
recreational city as it were, makes the car the ultimate hero, since there
has to be a means of transport linking the parts of city with each other.
However intelligently designed, a public commuter train network will
always be less flexible than the car. And so it was.

With the advent of the car, cities lose their human scale. "Today it is
really a mixed metaphor if, based on organic ideas, we speak of urban

Kranaufstellungen beim Hausbau – etwas, was selbst reale Sozialisten im Plattenbau später kopierten.

auto, straße, landschaft:
ein neues zeitalter beginnt

Weil die intellektuellen und kreativen Kräfte der Zwischenkriegszeit bereits auf wundersame Weise vernetzt sind – auch ohne Internet und Computer – wird das Gedankengut der Avantgarde sozusagen zum Schulwissen aller Planer und Ingenieure der Industriestaaten. Auf dem Kongress der CIAM (Congrès Internationaux d'Architecture Moderne) 1933 in Athen wird unter starker Mitarbeit von Le Corbusier die Charta von Athen verabschiedet, die den „autogerechten Städtebau" sanktionieren soll. Und das mindestens für die nächsten 100 Jahre. In der Charta stehen die Eckpunkte der Entwicklung wie in einem Gesetz, zum Beispiel zum Thema Verkehr: „Die Verkehrsstraßen müssen ihrem Charakter gemäß klassifiziert und entsprechend den Fahrzeugen und ihrer Geschwindigkeit gebaut werden". Oder: „Der Fußgänger muß andere Straßen als das Auto benutzen".[7] Die Aufteilung nach Funktionen, sozusagen in eine Wohn-, Geschäfts-, Arbeits- und Erholungsstadt, macht das Auto zum endgültigen Helden, denn es muss ein Verkehrsmittel geben, das die Teile der Stadt miteinander verbindet. Ein noch so intelligentes öffentliches Schnellbahnsystem wird dabei dem Auto immer an Flexibilität unterlegen sein. Und so kam es dann auch.

Mit dem Einzug des Autos verlieren die Städte ihren menschlichen Maßstab. „Es ist eigentlich ein schlechtes Bild, heute noch in Anlehnung an Organisches vom Städtewachstum zu sprechen", schrieb

growth", wrote Alexander Mitscherlich in his sweeping critique "The Inhospitality of the Modern City": "Cities are produced like cars".[8] He must have been thinking of de-individualization and uniformity. Despite all the fine theories, the measure of architectural practice now becomes a megalopolis, the best and clearest examples of whose development can be seen in the USA: in Los Angeles, for instance. It is made up of a "down town" with its bold office skyscrapers, which empties in the evenings, and a carpet known as an urban sprawl or plankton of suburbs and housing estates. The masses of cars, buses and lorries demand their tribute, and a unique aesthetic of traffic routes develops, made up of ten-lane highways, flyovers and underpasses, a pattern that is beautiful and impenetrable like spaghetti on a plate (see also *thoughts on the future of motortecture*). The only thing is that this is not pasta, but concrete. A monstrous devourer of space for which the car is responsible, an architectural drawing whose wondrous complexity can really only be deciphered from the air.

Usually, this was also how it came into being and was planned. Architectural models of cities have always been looked at from above. None of these urban monsters are conceived any longer from the perspective of the pedestrian and his perception, but from the "superior" structure of car users. For this reason, the logical consequence of this was also that there are really no longer any pedestrians in these cities: instead, isolated car pilots race through the large-scale iconography of the city, communicating inside their capsules via car radio or mobile telephone.

The pedestrian-free city finds its way right into the residential districts, planned without pavements. Cars come up the drive and disappear into

Alexander Mitscherlich in seiner Generalabrechnung „Die Unwirtlichkeit unserer Städte", und „Städte werden produziert wie Automobile".[8] Womit er Vermassung und Uniformität gemeint haben
muss. Statt aller schönen Theorie wird nun eine Megalopolis zum
Maßstab, am besten und klarsten in den USA entwickelt, zum Beispiel in Los Angeles. Sie setzt sich zusammen aus einer sich jeweils
in den Abendstunden entleerenden Downtown mit kühnen Büro-
hochhäusern und einem Teppich, den man einen Urban Sprawl oder
auch Plankton von Vororten und Wohnsiedlungen nennt. Die Massen
von Autos, Bussen und Lastwagen fordern ihren Tribut, es entwickelte sich eine eigene Ästhetik der Verkehrswege aus zehnspuri-
gen „Highways, Flyovers und Underpasses",[9] ein Muster schön und
undurchdringlich wie die Spaghetti auf dem Teller (vgl. *gedanken zur
zukunft der motortecture*). Nur dass es sich dabei nicht um
Teigwaren, sondern um Beton handelt. Ein riesiger Flächenfraß,
der dem Auto geschuldet ist, ein architektonisches Dessin, dessen
wundersame Komplexität eigentlich nur noch aus der Luft zu ent-
schlüsseln ist.

So sind Städte meistens auch entstanden und geplant worden. Archi-
tekturmodelle von Städten wurden immer nur von oben betrachtet;
alle diese Stadtmonster werden nicht mehr aus der Sicht des Fuß-
gängers und seiner Wahrnehmung erdacht, sondern aus der „überge-
ordneten" Struktur der Autonutzer heraus. Die logische Konsequenz
war deswegen auch, dass es Fußgänger in diesen Städten eigentlich
nicht mehr gibt. Vereinsamte Autopiloten jagen stattdessen durch die
großmaßstäbliche Ikonografie der Stadt, kommunizieren innerhalb
ihrer Kapsel über das Autoradio oder Mobiltelefon.

Dessin von wundersamer Komplexität: Autos, Straßen,
Flyovers und Underpasses prägen die Megalopolis.
A pattern of astonishing complexity: the megalopolis is the
product of cars, roads, flyovers and underpasses.

private garages. Cars that are left on the street overnight are broken into
now and then. Which is also a kind of automobile justice. The only true
counter-model may in fact be New York, which, to paraphrase Holland's
Rem Koolhaas, one of the 20th century's leading urbanists and town planning theorists, has become the expression of a "culture of congestion",[9]
although really it is a "culture of density" that has led to a simple two-
dimensional matrix being compacted and settled in the third dimension.
In the process, 2,028 blocks become 2,028 vertical cities within the city,
a miracle of compactness and one of the few urban models that has
come into being just so, without any theory. The car caused New York to
happen, but never fought or conquered it.

Over the last decades there have always been corrections and attempts
to impose order on the new city, and to see it as a structured road-
scape. This includes such ideological attempts as the *Reichsautobahn*
(imperial motorway). Of course, the National Socialists were disguising
their true intentions when they said it was all about automobility,
when really they meant rearmament. Even so, their holistic approach
was one that was kind to the landscape, built on compromise and co-
existence between the "road" and the countryside. The approach encom-
passed uniform turfed areas and borders, as well as curve radii and
cuttings that harmonized with the landscape. It included bridges, filling
stations and other buildings in a uniform philosophy which, although
nationalistically German, was also one of design.

The natural curve of the *Autobahn* was regarded more highly than making
the road as fast as possible: "A form that is technically good and correct
in every respect harmonizes easily with the environment, and for that
reason is in itself artistically good. Nature and art obey the same cosmic

Die fußgängerlose Stadt reicht bis hinein in die Wohnquartiere, die ohne Bürgersteige geplant sind. Die Autos fahren aufs Grundstück und verschwinden in den Privatgaragen. Autos, die auf den Straßen übernachten, werden bisweilen aufgebrochen. Auch eine Art von Autogerechtigkeit.

Als einziges wirkliches Gegenmodell mag dagegen nun wirklich New York gelten, das, nach einem der wichtigsten Urbanisten und Stadttheoretiker des 20. Jahrhunderts, dem Niederländer Rem Koolhaas, zum Ausdruck einer „Culture of Congestion"[10] geworden ist, einer „Kultur des Verkehrsstaus", wobei aber eigentlich aus einer „Kultur der Dichte" heraus ein einfach gerasterter zweidimensionaler Schachbrettgrundriss in der dritten Dimension verdichtet und besiedelt wird. 2028 Blöcke werden dabei zu 2028 vertikalen Städten in der Stadt, ein Wunder der Verdichtung und eines der wenigen Stadtmodelle, die ohne Theorie entstanden sind. Das Auto hat New York verursacht, aber nie bekämpft und besiegt.

Es gab im Laufe der letzten Jahrzehnte immer wieder Korrektive und Versuche, Ordnung in die neue Stadt zu bringen und sie als gestaltete Straßenlandschaft zu begreifen. Darunter so ideologische wie die Reichsautobahn. Dort hatten die Nationalsozialisten zwar die Tarnkappe aufgesetzt, als sie behaupteten, es ginge um Automobilität, wo sie Aufrüstung meinten. Ihr theoretischer Ansatz aber war ein landschaftsschonender und versöhnlicher, ein Miteinander der „Straße" und der Landschaft. Das reichte bis zu einheitlichen Grassoden und Bepflanzungen, zu landschaftsverträglichen Kurvenradien und Einschnitten, schloss Brücken, Tankstellen und Nebengebäude in eine, wenn auch deutschnationale, Gestaltungsphilosophie ein.

Vermutlich der einzige wirklich gelungene Versuch, ein Stadtmodell der Verdichtung und Urbanität mit dem Auto umzusetzen: Manhattan.
Probably the only truly successful attempt to put an urban model of compactness and urbanity into practice toegther with the car: Manhattan.

laws; in fact they do not contradict each other if they have not been somehow falsely understood",[10] was how someone on the staff of Fritz Todt, the master *Autobahn*-builder, once put it. However, this naturalistic, almost twee approach, which resulted in bridges being clad with local stone and routes being marked out with local church towers as orientation points, has as much to do with the later reality of mass motorization as a goldfish in its bowl has to do with the sea. Truly nothing remains of this originally meaningful quality of the *Reichsautobahn* in present-day Germany's motorway network, however.

As has already been mentioned, the car and its dynamism swept aside every filigree planning approach like a typhoon, including Bernhard Reichow's design for Sennestadt near Bielefeld, inspired stylistically by the veins of a leaf. For this dormitory town for Bielefeld, on the edge of the Teutoburg Forest, Reichow developed in the 1950s a system of feeder roads which had no crossroads, but only gentle loops and obtuse-angled junctions. A beautiful idea without any further consequences.

Or there is City Nord, a commercial quarter with roughly 30 large office blocks, built in the 1960s to relieve Hamburg's city centre. This satellite town's connections to Hamburg's commuter train network are only rudimentary, for it was designed as a green city in the name of the car, with generous multi-carriageway roads in a park with office buildings. All pedestrian movements were transferred to a second level (shades of Le Corbusier) of passages, stilts and ramps. Werner Hebebrand, Hamburg's chief architect at the time and the inventor of City Nord, commented: "Here, too, walking and driving are completely separate, there are no breakdowns, no congestion, no traffic restrictions, and there are large car-parking areas".[11]

Der natürliche Schwung der Autobahn war mehr wert als eine Geschwindigkeitsoptimierung: „Die in Hinblick auf alle Gesichtspunkte technisch gute und richtige Form fügt sich zwanglos der Natur wirklich ein und ist darum auch von selbst künstlerisch gut. Natur und Kunst gehorchen denselben kosmischen Gesetzen, eigentlich gibt es keinen Widerspruch zwischen ihnen, wenn sie nicht irgendwie falsch verstanden worden sind",[11] so fasste es einmal ein Mitarbeiter des damaligen Reichsautobahn-Baumeisters Fritz Todt zusammen. Der naturalistische bis beinahe niedliche Denkansatz, der dazu führte, dass Brücken mit Steinen aus der Umgebung verkleidet und die Trassen auf örtliche Kirchtürme orientiert wurden, hat allerdings mit der späteren Wirklichkeit der Massenmotorisierung so viel zu tun wie der Goldfisch im Glas mit dem Meer. Von der ursprünglich sinnstiftenden Qualität der Reichsautobahnen ist im bundesdeutschen Autobahnnetz nun wirklich nichts übrig geblieben.

Es wurde schon erwähnt: Das Auto mit seiner Dynamik hat alle fein konstruierten planerischen Ansätze hinweggefegt wie ein Taifun, unter anderen auch den stilistisch dem Adersystem eines Baumblatts nachempfundenen Entwurf von Bernhard Reichow für die Sennestadt bei Bielefeld. Der hatte in den fünfziger Jahren für diese Wohnstadt der Stadt Bielefeld am Teutoburger Wald ein Straßenerschließungsbild entwickelt, bei dem es keine Kreuzungen, sondern nur sanfte Schleifen und Einmündungen gibt. Eine schöne Idee ohne weitere Folgen.

Oder die Hamburger Geschäftsstadt City Nord, die in den sechziger Jahren zur Entlastung der Hamburger Innenstadt gebaut wurde

Schnellstraße mit Respekt vor der Landschaft:
das ursprüngliche Konzept der Reichsautobahn.
Expressway that respects the countryside:
the original concept behind the Reichsautobahn.

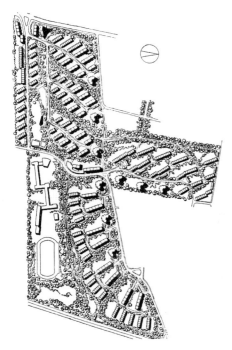

Schema eines autogerechten Hamburger Stadtteils
aus den fünfziger Jahren von Bernhard Reichow,
dem Planer der Sennestadt.
Plan of a car-scaled Hamburg district from the 1950s by
Bernhard Reichow, the planner of Sennestadt

City Nord has remained an alien element in Hamburg's cityscape, though it still works even today. Its distribution and scaling of traffic areas is meanwhile regarded as false, and the model of a self-contained office quarter has failed because it has failed to develop true urbanity. One further footnote, albeit an amusing one, is provided by the urban utopias of the British Archigram group in the 1960s. Their daring utopias have nothing to do with buildings and architecture. They are like perfect science fiction, steam-rolling all conventions, whether building or car. Archigram's apotheosis, invented by its founder Ron Herron, was the "Walking City" project of 1964 – an automobile city, as it were, moving on legs to take its inhabitants to the desert or Greenland, as their fancy took them. Even in 1964, however, it was clear that "despite their monstrous technological perfection, they were more like the helpless relics of an epoch that still optimistically believed in technological progress".[12]

the american strip:
dynamism and persuasion

The American city, which has gone furthest in accommodating the car, has done little to foster urbanity. What it has done is to develop an iconography of its own, in terms of where each town's main facilities are located. In the meanwhile, there now exists there a very complicated interplay between car and architecture. Both are part of everyday (American) life, both are everyday heroes, without having to be theatrical, even though they are so. The best example of this is the American "strip" or "Main Street", which no longer has anything in common with the good old European High Street. Everything here is devoted to the car and its speed: the curves of entrances, advertising, everything people

und etwa 30 Bürogroßbauten aufnahm. Diese Satellitenstadt ist nur dürftig an das Hamburger Schnellbahnnetz angebunden, sie wurde als grüne Stadt im Namen des Autos entworfen, mit großzügigen mehrspurigen Straßen in einem Park mit Bürohäusern. Der gesamte Fußgängerverkehr wurde in eine zweite Ebene (siehe Le Corbusier) auf Passagen, Stelzen und Rampen gelegt. Werner Hebebrand, der damalige Hamburger Oberbaudirektor und Erfinder der City Nord schrieb dazu: „Auch hier völlige Trennung von Fahren und Gehen, keine Betriebsstörungen, keine Verstopfungen, keine Verkehrsregelungen, große Abstellflächen für Kraftfahrzeuge".[12] Die City Nord ist im Stadtbild Hamburgs ein Fremdkörper geblieben, auch wenn sie bis heute funktioniert. Ihre Verteilung und Dimensionierung von Verkehrsflächen gilt inzwischen als falsch, das Modell der autarken Bürostadt ist insofern gescheitert, als sie keine echte Urbanität entwickelt hat.

Ebenso eine Fußnote, allerdings eine amüsante, blieben die Stadtutopien der britischen Gruppe Archigram in den sechziger Jahren. Deren tollkühne Entwürfe haben mit Gebäuden und Architektur nichts zu tun. Sie gleichen einer perfekten Science-Fiction und überrollen alle Konventionen, ob Haus oder Auto. Archigrams Höhepunkt, erfunden durch ihren Gründer Ron Herron, war 1964 das Projekt „Walking City", eine automobile Stadt sozusagen, die sich auf Makro-Stelzen mit ihren Bewohnern nach deren Gusto wahlweise in die Wüste oder nach Grönland bewegt. Allerdings war schon damals deutlich, dass diese Utopien „trotz ihrer furchterregenden technologischen Vollkommenheit eher wie die hilflosen Relikte einer vom technischen Fortschrittsoptimismus geprägten Epoche"[13] wirken.

Zur Entlastung der Hamburger Innenstadt geplant: City Nord, eine Planung „im Namen des Autos" (sechziger Jahre, Konzept: Werner Hebebrand).

Planned to relieve Hamburg's inner city: City Nord, an example of town planning "for the car's sake" (1960s. Planner: Werner Hebebrand).

need, whether petrol station or brothel, stands in a row along the highway.

In the 1960s, the American architects Robert Venturi and Denise Scott Brown decoded the semantics of American street architecture in an architectural cult book ("Learning from Las Vegas"[13]), deriving from the culture of advertising along the highway the idea of the "decorated shed", and with it a new architectural language. Because this resulted in a fragmentary collage of quotations from the history of architecture and of trivial forms from everyday automotive life, Venturi is in some ways the father of American postmodernism, an architectural language that owes its origins to the motor car.

While some may appreciate the irony that arises from the vernacular and trivial and that runs counter to and questions high modernism, and others may find that the first true street culture of the 20th century is shabby and improper, it cannot be denied that carved lattice work, caryatids and atlantes are no longer the measure of all things. Architecture is no longer something to be digested by strolling pedestrians, and the tactile quality of the building materials no longer plays any role in the sense that people want to stroke them lovingly. Here, everything is geared to rapid perception, to effect and signposting from a distance, be it via extra large signs or coloured guidance systems. Here, the sign at the Shell filling station is just as important as the freeway number on a road sign. This city is geared to the car, including the perception of drivers passing by. Cars and their drivers are players on a street stage, whose backdrops do not disturb the flow of vehicles.

der amerikanische strip:
dynamik und fernwirkung

Die amerikanische Stadt trug wenig zur Förderung der Urbanität bei, obwohl sie sich am konsequentesten dem Auto unterworfen hat. Aber sie hat eine eigenständige Ikonografie entwickelt, die festschreibt, wo die zentralen Versorgungseinrichtungen der jeweiligen Ortschaften liegen. Inzwischen gibt es dort eine sehr diffizile Wechselwirkung zwischen Auto und Architektur. Beide gehören zum (amerikanischen) Alltag, beide sind Helden des Alltags, ohne dass sie theatralisch sein müssen. Sie sind es aber. Das beste Beispiel ist hier der amerikanische „Strip", die „Main Street", die mit der guten alten deutschen Bahnhofstraße nichts mehr gemein hat. Alles ist hier auf das Auto und seine Geschwindigkeit ausgerichtet: Einfahrtsradien, Werbung, alles, was der Mensch braucht, ob Petrol Station oder Puff, alles ist hintereinander am Highway aufgereiht. Die amerikanischen Architekten Robert Venturi und Denise Scott Brown haben in den sechziger Jahren mit einem architektonischen Kultbuch („Lernen von Las Vegas"[14]) die Grammatik dieser amerikanischen Straßenarchitektur dekodiert, haben aus der Kultur der Werbung am Highway den „dekorierten Schuppen" und damit eine neue Architektursprache abgeleitet. Weil sich daraus eine fragmentarische Collage aus Zitaten aus der Baugeschichte und trivialen Formen des Autoalltags ergab, ist Venturi auch so etwas wie der Vater der amerikanischen Postmoderne, einer Architektursprache, die weitgehend dem Auto geschuldet ist.

Mögen die einen die Ironie schätzen, die sich aus Banalitäten und Trivialitäten ergibt und die kultivierte Baugeschichte konterkariert

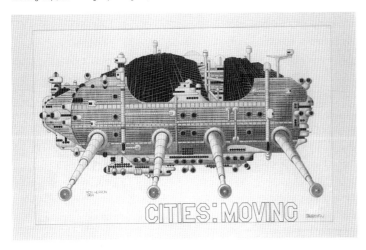

Stadtutopie als Science-Fiction:
Walking City (1964, Entwurf Archigram).
Urban utopia as science fiction:
Walking City (1964. Design by Archigram).

Von Venturi gelernt: vorbildliche Architektur für einen deutschen Autostrip in Hamburg (Showroom „Car and Driver", 1991, Architekten: Hadi Teherani und Wolfgang Raderschall).
Venturi showed the way: excellent architecture for a German strip in Hamburg ("Car and Driver" showroom, 1991. Architects: Hadi Teherani and Wolfgang Raderschall).

Dutch architects such as the Rotterdam MVRDV group have taken this topic further, even though their architectural understanding of it is different, because it is rational and Calvinist. Their prime aim is to stop cities' tendency to sprawl; in other words, to compact human settlement and expand natural spaces. This in turn results in a special architecture "for the edge" – for the limits of settlement, which are always something that drivers on expressways are especially on the look-out for.

in a stranglehold –
or with a future after all?

That brings us to the present day, at the beginning of the 21st century. Compared with classical modernism, the view of architecture under the sign of the motor car has changed and moved on, because computer-based drafting and design work has opened up new fields. Today's young architectural talent is concerned with the chance fixed image, for example, the momentary record of dynamic movement. The new heroes of a biomorphic architecture are Greg Lynn, Hani Raschid or Lars Spuybroek. A young German architect called Bernhard Franken has designed a dynaform as a mobile pavilion for the BMW group, or, to be more precise, has used a computer to generate its design.

Now, in the 21st century, architects are intoxicated by the dynamic, flowing form of a speeding car. Or, as Gerhard Matzig put it in the *Süddeutsche Zeitung*, as experts of immobility they want to dream of acceleration – instead of "Peter the rock, they want to be Schumacher the racing car".[14]

und in Frage stellt, mögen andere hingegen diese erste richtige Straßenkultur des 20. Jahrhunderts unseriös und unwürdig finden, in jedem Fall gilt: Hier setzen keine Fachwerkschnitzereien, keine Atlanten und Karyatiden den Maßstab, der von schlendernden Fußgängern verarbeitet würde, hier spielt auch die Haptik des Baumaterials keine Rolle, weil Menschen es liebevoll streicheln wollten. Hier ist alles auf schnelle Wahrnehmung abgestellt, auf Fernwirkung und Fernlenkung, sei es durch XL-Schriftgrößen oder farbige Leitsysteme; hier ist das Markenschild der Shell-Tankstelle ebenso wichtig wie die Autobahnnummer auf der Hinweistafel. Hier ist die Stadt autogerecht, auch dort, wo es um die Wahrnehmung der Autopassanten geht. Die Autos und ihre Fahrer sind die Akteure auf einer Straßen-Bühne, deren Kulissen den Fluss der Fahrzeuge nicht stören.

Dieses Thema wird heute – wenn auch mit einem konsequent veränderten architektonischen, weil rationalistischen, kalvinistischen Verständnis – von niederländischen Architekten wie der Rotterdamer Gruppe MVRDV weitergedacht. Sie will vor allem die zersiedelnden Tendenzen der Städte stoppen. Daraus ergibt sich wiederum auch eine spezielle Architektur für die Ränder der Besiedlung, „for the edge", die immer unter der besonderen Wahrnehmung der Autofahrer auf Schnellstraßen stehen.

**im würgegriff –
oder doch mit zukunft?**

Womit wir in der Gegenwart, am Anfang des 21. Jahrhunderts, angekommen wären. Die Auffassung von Architektur im Zeichen des Autos hat sich im Vergleich zur klassischen Moderne verändert

Bauen für den Rand und die Wahrnehmung des Autofahrers: Entwürfe von MVRDV (Niederlande) für einen Gewerbepark in Eindhoven (1998).
Building for the edge and for the perception of the car driver: designs by MVRDV (Netherlands) for a business park in Eindhoven (1998).

Today's highly sophisticated software, adapted from the aviation and car manufacturing industry, is literally causing the borders between car and house construction to blur. In this small design field, the car and architecture have become reconciled.

The end of this process is the car-scaled town without cars (see also *autostadt*) and architects learning from car engineers, but there is another lesson here. At the end of the 20th century, rivalries and opposing standpoints have been abandoned, and we can speak of a kind of MotorTecture.[15] A married couple have been forced to decide to grow old together until – who? – them do part.

What are the standpoints? A city built for cars does not exist, perhaps one day there will be one that is human-scaled and that integrates the car. At present, it is still the other way around. Economically and psychologically, doing without the car as a means of mass transport is inconceivable. And so, we learn to live with it. And we hope that this learning to live with it is more than an unthinking idolatry of the car, in which architecture plays the role of a golden altar.

In the automobile society, we hope that architecture is not just a backdrop for mobile observers, but that is fun and meaningful; as housing, i.e. as a building in which to park and stow away, and as a companion along the route, whether as an architecturally striking bridge or well-designed filling station. This book presents the first proud batch of examples.

But we want more than this. The interface between immobile and mobile, between house or garage and car or movement has still not been mastered architecturally. For many decades, of course, there have

und weiterentwickelt, weil die Entwurfs- und Konstruktionsarbeit mit dem Computer neue Felder erschlossen hat. Jetzt steht beispielsweise das zufällige Vexierbild, die Momentaufnahme der dynamischen Bewegung, auf dem Stundenplan einer frischen Architektengeneration. Die neuen Helden einer biomorphen Architektur heißen Greg Lynn, Hani Raschid oder Lars Spuybroek. Ein junger deutscher Architekt namens Bernhard Franken entwirft eine Dynaform als Wanderpavillon für den BMW-Konzern, genauer gesagt, er lässt sie vom Computer generieren.

Jetzt, schon im 21. Jahrhundert, sind Architekten von der dynamischen, der Fließform eines dahinrasenden Autos berauscht, wollen sie, wie Gerhard Matzig in der *Süddeutschen Zeitung* schrieb, als Experten der Immobilität vom Gasgeben träumen, statt wie „Petrus der Fels wie Schumi der Bolide sein".[15] Heute bringt eine gnadenlos raffinierte Software, die aus der Flugzeug- oder eben Autoindustrie übernommen wird, die Grenzen zwischen Auto- und Hauskonstruktion buchstäblich zum Fließen. Dort, in diesem kleinen Design-Bereich, ist die Versöhnung zwischen Auto und Architektur vollzogen. Am Ende stehen eine autogerechte Stadt ohne Autos (vgl. *autostadt*) und das Lernen der Architekten von den Autoingenieuren, aber es gibt auch noch eine andere Erkenntnis. Zum Ende des 20. Jahrhunderts sind Konkurrenzen und Gegenpositionen aufgegeben, man kann von einer Art MotorTecture[16] reden. Zwei Ehepartner haben zwangsweise beschlossen, miteinander uralt zu werden, bis dass – wer? – sie scheidet.

Was sind die Positionen? Die autogerechte Stadt gibt es nicht, vielleicht irgendwann einmal jene, die dem Menschen gerecht wird

been drive-in buildings. But what do they look like? Like 1970s underground stations, ashamed to show their faces. Or the urban fringes of expressways – their design is geared at best to sentimental American archetypes, more retro than forward-looking.

The narrative of modern architecture will no longer be able to ignore the requirements of mobility, its form of expression and its quite specific needs. The reverse is also true, of course. MotorTecture is still in its infancy, and will not end until, true to its own dynamism, it drives to its own death. But meanwhile, of course, we have also learned that it is possible to drive defensively.

und das Auto integriert. Zurzeit ist es noch andersherum. Ökonomisch und psychologisch ist der Verzicht auf Autos als Massenverkehrsmittel undenkbar. Wir arrangieren uns also. Und wir hoffen, dass dieses Arrangement mehr ist als eine unreflektierte Götzenanbetung des Autos, bei der die Architektur die Rolle des goldenen Altars spielt.

Wir hoffen, dass Architektur in der Autogesellschaft nicht nur Kulisse der in Bewegung befindlichen Beobachter ist, sondern ihren Sinn hat und Spaß macht; sei es als Behausung, sprich Haus zum Parken und Wegstellen, als Begleiter der Trassen, sei es als architektonisch markante Brücke und schöne Tankstelle. Eine erste stolze Schar von Beispielen liefert dieses Buch. Aber wir wollen noch mehr. Noch immer ist die Schnittstelle zwischen dem Immobilen und Mobilen, zwischen Haus oder Garage und Auto oder Bewegung architektonisch nicht bewältigt. Es gibt zwar seit vielen Jahrzehnten Drive-In-Gebäude. Nur – wie sehen sie aus? Wie verschämt versteckte U-Bahn-Stationen der siebziger Jahre. Die urbanen Ränder an Schnellstraßen – ihre Gestaltung folgt allenfalls sentimentalen Urbildern aus Amerika und ist mehr Retro denn Zukunft.

Die Geschichte der modernen Architektur wird nicht länger die Anforderungen der Mobilität, den Ausdruck ihrer ganz spezifischen Bedürfnisse ignorieren können. Umgekehrt natürlich auch nicht. Die MotorTecture steht erst an ihrem Anfang und sie endet erst dann, wenn sie sich ihrer eigenen Dynamik gehorchend zu Tode fährt. Aber es gibt ja auch die Möglichkeit einer defensiven Fahrweise.

1 Mateo Kries
  „Vom Auto zur Massenmotorisierung"
  in: Volker Albus, Mateo Kries,
  Alexander von Vegesack (Hrsg.)
  Automobility. Was uns bewegt
  Weil am Rhein 1999, S. 194.
2 Zur Vertiefung empfohlen:
  Gert Kähler
  Architektur als Symbolverfall.
  Das Dampfermotiv in der Baukunst
  Braunschweig 1981.
3 Renzo Piano
  Mein Architektur-Logbuch
  Ostfildern-Ruit 1997, S. 92.
4 Heinrich Klotz, John W. Cook
  Architektur im Widerspruch.
  Bauen in den USA von Mies van der
  Rohe bis Andy Warhol
  Zürich 1974, S. 144.
5 Kenneth Frampton
  Die Architektur der Moderne.
  Eine kritische Baugeschichte
  Stuttgart 1983, S. 133.
6 ebd., S. 135.
7 Ernesto Grassi (Hrsg.)
  Le Corbusier. Charta von Athen
  Reinbek bei Hamburg 1962, S. 107 ff.
8 Alexander Mitscherlich
  Die Unwirtlichkeit unserer Städte.
  Anstiftung zum Unfrieden
  Frankfurt 1969, S. 33.
9 Diese englischen Begriffe verdeutlichen,
  dass Hochstraßen, Überflieger und

Tunnel das Straßensystem dreidimensional gemacht haben.
10 Rem Koolhaas
  Delirious New York.
  Ein retroaktives Manifest für Manhattan
  Aachen 1999, S. 124.
11 Erhard Schütz, Eckhard Gruber
  Mythos Reichsautobahn.
  Bau und Inszenierung der
  „Straßen des Führers" 1933–1941
  Berlin 1996, S. 122 ff.
12 Werner Hebebrand in der Ausstellung
  City Nord, Hamburg 2000.
13 Heinrich Klotz
  Vision der Moderne.
  Das Prinzip Konstruktion
  München 1986, S. 316.
14 Zur Vertiefung empfohlen:
  Steven Izenour, Denise Scott Brown,
  Robert Venturi *Lernen von Las Vegas*
  Braunschweig, Wiesbaden 1981.
15 Gerhard Matzig *„Auto, Motor und Gott"*
  in: Süddeutsche Zeitung vom 21.8.2002.
16 MotorTecture, Carchitecture, Motor-
  Tektur sind Kunstbegriffe und noch
  nicht etabliert. Sie versuchen die Beziehung zwischen dem Auto und der
  Architektur in einem Wort auszudrücken.
  „Carchitecture" ist auch der Titel eines
  Buches des britischen Journalisten
  Jonathan Bell (siehe Literaturverzeichnis). Es sei ausdrücklich zur Vertiefung
  des Themas empfohlen.

All references in the footnotes refer to
the respective German editions and texts.
1 Mateo Kries
  "Vom Auto zur Massenmotorisierung",
  in: Volker Albus, Mateo Kries,
  Alexander von Vegesack (Eds)
  Automobility – Was uns bewegt
  Weil am Rhein 1999, p. 194.
2 Recommended:
  Gert Kähler
  Architektur als Symbolverfall.
  Das Dampfermotiv in der Baukunst
  Braunschweig 1981.
3 Renzo Piano
  Mein Architektur-Logbuch
  Ostfildern-Ruit 1997, p. 92.
4 Heinrich Klotz, John W. Cook
  Architektur im Widerspruch.
  Bauen in den USA von Mies van der
  Rohe bis Andy Warhol
  Zurich 1974, p. 144.
5 Kenneth Frampton
  Die Architektur der Moderne.
  Eine kritische Baugeschichte
  Stuttgart 1983, p. 133.
6 ibid., p. 135.
7 Ernesto Grassi (Ed.)
  Le Corbusier. Charta von Athen
  Reinbek bei Hamburg 1962, p. 107 ff.
8 Alexander Mitscherlich
  Die Unwirtlichkeit unserer Städte.
  Anstiftung zum Unfrieden
  Frankfurt 1969, p. 33.

9 Rem Koolhaas
  Delirious New York.
  Ein retroaktives Manifest für Manhattan
  Aachen 1999, p. 124.
10 Erhard Schütz, Eckhard Gruber
  Mythos Reichsautobahn.
  Bau und Inszenierung der
  "Straßen des Führers" 1933–1941
  Berlin 1996, p. 122 ff.
11 Werner Hebebrand in the exhibition
  City Nord, Hamburg 2000.
12 Heinrich Klotz
  Vision der Moderne.
  Das Prinzip Konstruktion
  Munich 1986, p. 316.
13 Recommended:
  Steven Izenour, Denise Scott Brown,
  Robert Venturi
  Lernen von Las Vegas
  Braunschweig, Wiesbaden 1981.
14 Gerhard Matzig *"Auto, Motor und Gott"*
  in: Süddeutsche Zeitung 21.8.2002.
15 Terms such as motortecture and
  carchitecture are artificial, and have
  yet to become established. They are an
  attempt to express the relationship
  between the car and architecture in one
  word. "Carchitecture" is also the title
  of a book by the British journalist
  Jonathan Bell (see bibliography). The
  book can be warmly recommended to
  any reader wishing to find out more
  about this topic.

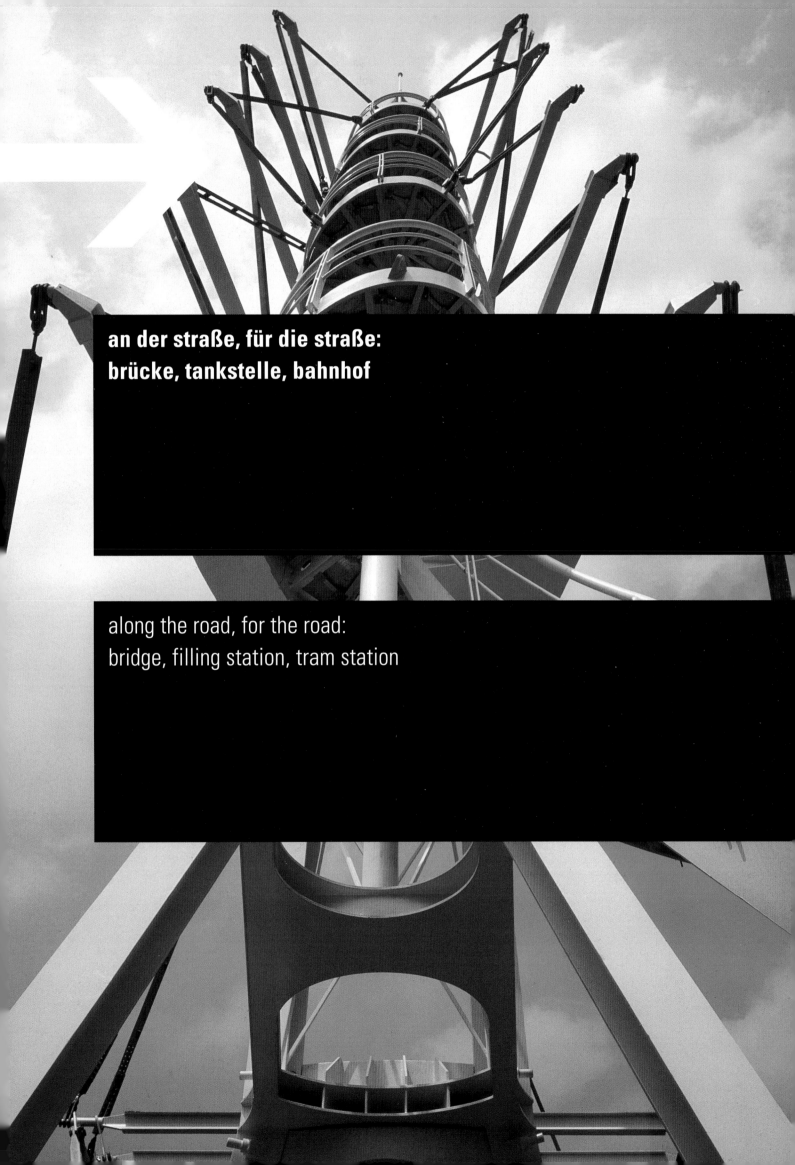

an der straße, für die straße:
brücke, tankstelle, bahnhof

along the road, for the road:
bridge, filling station, tram station

wege über das meer: normandiebrücke

Das Auto ist nur so gut wie die Spur, auf der es fährt. Denn das Auto kann weder fliegen noch Berge versetzen – die Automobilität hat eben technische Grenzen. Allerdings sind Ingenieurbaukunst und Architektur inzwischen in der Lage, Berge raffiniert zu untertunneln und andere Hindernisse kühn zu überbrücken. Die Nutzbarmachung von Architektur und Technik durch das Auto wird besonders deutlich an Brücken, denn inzwischen werden Meere, zumindest Teile von ihnen, mit gigantischen Brückenschlägen überspannt. Auf diese Weise kann vielleicht schon bald sogar Sizilien mit dem italienischen Festland über die Straße von Messina verbunden werden.

Drei neue europäische Meeresbrücken aus den letzten Jahren stehen stellvertretend für diese Entwicklung: Die Öresundbrücke zwischen Kopenhagen und Malmö schließt eine der letzten Straßen-Lücken zwischen Gibraltar und dem Nordkap. Ihre Nachbarin über den Großen Belt gehört bei einer freien Spannweite von 1.624 m zu den längsten Brücken der Welt. Das dritte Beispiel, die Normandiebrücke bei Le Havre, nennt einer der wichtigsten Brückenkonstrukteure der Welt, Jörg Schlaich (s. auch S. 36 ff, 58), bautechnisch „bedeutungs- und phantasievoll".

Normandiebrücke
Autobahnbrücke zwischen Le Havre und Calvados über die Seine

Bauzeit	1989 – 1995
Länge	2.141 m
Hauptspannweite	856 m
Konstruktion	Schrägseilbrücke
Höhe der Pylonen	214,77 m
Architekten	François Doyelle, Charles Lavigne, Vanves
Ingenieur	Michel Virlogeux, Bagneux

routes across the sea: pont de normandie

Cars are only as good as the roads they drive along. For the car can neither fly nor move mountains. In other words, there are technical limits to automobility. Having said that, skilful engineers and architects have meanwhile learnt how to drive tunnels through mountains and to build bold bridges over other obstacles. The way the car makes use of architecture and technology is especially obvious in the case of bridges, for meanwhile seas, or least parts of them, are literally being crossed by huge bridges linking opposite shores. In this way, even Sicily may soon be linked with the Italian mainland across the Straits of Messina.

Three new European sea bridges from recent years illustrate this adventurous, bold development: the Öresund Bridge between Copenhagen and Malmö closes one of the last gaps on the road from Gibraltar to the North Cape. With a free span of 1,624 metres, its neighbour over the Great Belt is one of the longest bridges in the world. Our third example, the Pont de Normandie near Le Havre, has been described by Jörg Schlaich, one of the world's leading bridge builders (see also p. 36 ff, 58), as structurally "significant and imaginative".

Pont de Normandie
Motorway bridge between Le Havre and Calvados, across the Seine

Constructed	1989 – 1995
Length	2,141 m
Main span	856 m
Structure	Cable-stayed bridge
Pylon height	214.77 m
Architects	François Doyelle, Charles Lavigne, Vanves
Engineer	Michel Virlogeux, Bagneux

Über die weite Seinemündung bahnt sich die Normandiebrücke schwungvoll ihren Weg, gehalten durch zwei gut 200 m hohe Pylonen in der Form eines übergroßen „A".
The Pont de Normandie dynamically carves out a route across the wide estuary of the Seine, supported by two pylons shaped like gigantic "A"s, each more than 200 metres high.

wege über das meer:
brücke über den großen belt und öresundbrücke

Thema in diesem Zusammenhang sind nicht ingenieurtechnische Wundertaten, sondern die perfekte Verbindung aus Technik und Ästhetik, dort, wo Architektur und Auto sich sehr nahe kommen. Da heute Brücken dieser Dimension als Zeichen gesetzt und gesehen werden, steht ihr architektonischer Ausdruck zwangsläufig im Mittelpunkt. Bei Schrägseilbrücken werden dabei Gestaltung und Dimensionierung der Pylonen zum Charakteristikum der Gesamtveranstaltung. In der Normandie paart sich die schwungvolle Straße mit einer eleganten Überhöhung der Pylonen, im Öresund grüßt ein stolzer Vierzack. Und die zweitlängste Hängebrücke der Welt über den Belt wartet mit dem ewig jungen Abbild der Golden Gate Bridge in San Francisco auf. Was in der Stadt nicht gelingt, ist hier über dem Meer Standard: die Versöhnung und gemeinschaftliche Aktivität von Baukunst und Auto.

routes across the sea:
great belt bridge and öresund bridge

The topic here is not one of astounding engineering feats, but of the perfect combination of technology and aesthetics at the point where architecture and the car come very close: as bridges of this size nowadays set an example and are regarded as such, attention is bound to be focussed on their architectural expression. In the case of cable-stayed bridges, the design and size of the pylons are the distinctive feature of the overall bridge "event". In Normandy, the dynamic line of the motorway is combined with the elegantly exaggerated height of its pylons, in Öresund a proud four-pronged spear greets the spectator. And the world's second longest suspension bridge over the Great Belt is the forever young likeness of San Francisco's Golden Gate Bridge. Here, over the sea, a standard has been achieved where cities have failed: the reconciliation between and joint activity of architecture and the car.

Brücke über den Großen Belt
Autobahnbrücke zwischen den dänischen Inseln Fünen und Seeland (Jütland–Kopenhagen)

Bauzeit	1991–1998
Länge	6.800 m
Hauptspannweite	1.624 m
Konstruktion	Hängebrücke
Höhe der Pylonen	254 m
Lichte Höhe	57 m
Architekten	Dissing + Weitling, Kopenhagen
Ingenieure	COWI A/S, Lyngby

Great Belt Bridge
Motorway bridge between the Danish islands Fünen and Seeland (Jutland–Copenhagen)

Constructed	1991–1998
Length	6,800 m
Main span	1,624 m
Structure	Suspension bridge
Pylon height	254 m
Clear height	57 m
Architects	Dissing + Weitling, Copenhagen
Engineers	COWI A/S, Lyngby

Brücke über den Belt: ein europäisches Golden Gate
Bridge over the Belt: a European Golden Gate

Stolzer Vierzack: Öresundbrücke
Proud four-pronged spear: Öresund Bridge

Öresund-Brücke
Autobahn- und Eisenbahnbrücke zwischen Malmö und Kopenhagen über den Öresund

Bauzeit	1996–2000
Länge	7.845 m
Konstruktion	Schrägseilbrücke
Höhe der Pylonen	203,5 m
Größte Spannweite	1.092 m
Lichte Höhe	57 m
Architekt	Georg Rotne, Kopenhagen
Ingenieure	Gimsing & Madsen, Hirtshals, ISC, Kopenhagen, Ove Arup + Partners, London, Setec TPI, Paris

Öresund Bridge
Motorway and railway bridge between Malmö and Copenhagen, across Öresund

Constructed	1996–2000
Length	7,845 m
Structure	Cable-stayed bridge
Pylon height	203.5 m
Largest span	1,092 m
Clear height	57 m
Architect	Georg Rotne, Copenhagen
Engineers	Gimsing & Madsen, Hirtshals, ISC, Copenhagen, Ove Arup + Partners, London, Setec TPI, Paris

klipp, klapp und weg: klappbrücke kiel-hörn

Zweifeldklappbrücken, so der Architekt Volkwin Marg, gehören im Schiffsbau zum Standard. Dreifeldklappbrücken für den Straßenverkehr, die nach demselben Prinzip arbeiten, sind eine Neuheit, und so dauerte es auch ein bisschen länger bis zur Fertigstellung der neuen Verbindung zwischen der dicht besiedelten Kieler Innenfördeseite und neuen Wohngebieten, wo einst maritime Industrie ihr Zuhause hatte. Am Ende stand ein Produkt, das eine Durchfahrtsbreite von 22 m zulässt, weil es einem raffinierten Zug-Klapp-Mechanismus gehorcht: die Felder falten sich beim Hochziehen wie eine Ziehharmonika zusammen. Einfach und trotzdem markant sollte das kleine technische Wunder sein. Das ist unter anderem auch deswegen gelungen, weil die unterschiedlichen Teile der Konstruktion in verschiedenen leuchtenden Farben angestrichen wurden und auf diese Weise die Zug- und Faltbewegungen zusätzlich markiert sind. Schade, dass diese Brücke eigentlich nicht dem Auto, sondern nur Fahrradfahrern und Fußgängern zur Verfügung steht ...

fold up and away: bridge across the hörn (kiel firth)

According to the architect Volkwin Marg, double-leaf bascule bridges are normal in shipbuilding. Three-leaf bascule bridges for road traffic, which work according to the same principle, are something new, and so it took a little longer until the new link was completed between the densely settled inland side of Kiel Firth and new residential areas located on the former site of maritime industries. The end result was a product that allows ships as wide as 22 metres to pass, because it obeys a sophisticated draw-fold mechanism, its leaves folding together like an accordion when it is raised. "Simple yet striking" was the brief for this minor technological wonder. One reason why it fulfilled this successfully was that the different parts of the structure were painted in different luminous colours, in this way additionally emphasizing the draw and fold movements. The only pity is that this bridge is not really open to motor vehicles, but only to cyclists and pedestrians ...

Die Klappbrücke besteht aus drei Feldern.
The bascule bridge comprises three leaves.

Phasen des Aufklappens zur Vorbereitung einer Schiffsdurchfahrt.
The bridge is folded to allow a ship to pass.

**Konstruktionsdetails. Der farbig unterschiedliche Anstrich
verdeutlicht die verschiedenen konstruktiven Aufgaben
des Brückenmechanismus.**
Design details. Painting the bridge in different colours emphasizes
the different constructional tasks of the bridge mechanism.

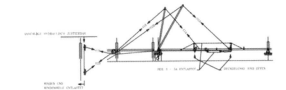

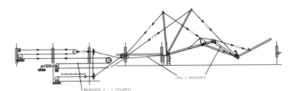

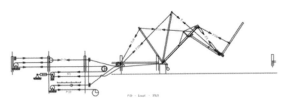

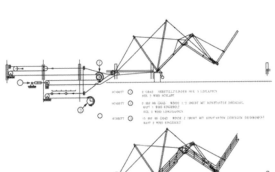

Klappbrücke Kiel-Hörn

Schrägseilbrücke mit zwei zu jeweils zwei Bügeln verbundenen beweglichen Pylonen und Dreifeldklappmechanismus

Bauzeit	**1996–1997**
Länge	**116 m, davon ca. 25 m als Faltbrücke ausgebildet**
Architekt	**Volkwin Marg (von Gerkan, Marg und Partner), Hamburg**
Ingenieure	**Schlaich, Bergermann und Partner, Stuttgart**

Bascule bridge, Kiel-Hörn

Cable-stayed bridge with two movable pylons, each connected to two stays, and a three-leaf folding mechanism

Constructed	1996–1997
Length	116 m, 25 m of which is designed as a drawbridge
Architect	Volkwin Marg (von Gerkan, Marg und Partner), Hamburg
Engineers	Schlaich, Bergermann und Partner, Stuttgart

Erläuterungsskizzen zur Brückenkonstruktion
Explanatory sketches of the bridge design

umkehrschluss: sunnibergbrücke bei klosters

Vom Brückenbauer Christian Menn aus der Schweiz ist der Satz überliefert: „Eigentlich hat sich seit der Golden Gate Bridge (1933–37) nichts Neues mehr beim Brückenbau ergeben". Menn tritt mit seiner 526 m langen Sunnibergbrücke bei Klosters, Kanton Graubünden, den Versuch an, seine eigene These zu widerlegen. Dabei bedient er sich des optischen Tricks, dass er die üblicherweise in den Himmel ragenden Pylonen nicht in Form eines „A", sondern in der Form eines „V" in den Boden steckt: „Wir wollen nicht eine Brücke durch den Wald bauen, sondern den Eindruck erwecken, dass sie sozusagen aus dem Wald herauswächst!" Außerdem kann diese Schrägseilbrücke nicht zu einem Fremdkörper in der Landschaft werden, da das Straßenband fast so dünn wie eine Papierbahn wirkt. Ein „technisches Meisterwerk" nannte sie die renommierte Architekturzeitschrift *Domus*.

Bauzeit	1996–1998
Länge	526 m
Höhe der Pylonen	max. 77 m
Ingenieur	Christian Menn, Chur
Beratender Architekt	Andrea Deplazes, Chur

reverse conclusion: sunniberg bridge near klosters

Swiss bridge-builder Christian Menn is on record as having stated: "In reality, there has been nothing new in bridge-building since the Golden Gate Bridge (1933–37)". With his 526 metre long Sunniberg Bridge near Klosters, in the Swiss canton Graubünden, Menn attempted to refute his own thesis. His attempt uses an optical trick: instead of setting his pylons in the ground in the form of an "A" rising into the sky, he uses a "V" form. "We don't want to build a bridge through the forest, but to arouse the impression that it is growing out of the forest, as it were". And besides, this cable-stayed bridge cannot be an alien element in its setting, as its deck has been designed to be almost as thin as a strip of paper. The highly regarded architectural periodical *Domus* called it a "technological masterpiece".

Constructed	1996–1998
Length	526 m
Pylon height	max. 77 m
Engineer	Christian Menn, Chur
Consulting Architect	Andrea Deplazes, Chur

Im Alpenland Schweiz gehören Lawinenschutz, Tunnel- und Brückenbauten zum Architektenalltag. Projekte wie die Sunnibergbrücke beweisen, dass auch Technikbauwerke für den Autofahrer mit Esprit und Eleganz in die gewaltige Landschaft eingefügt werden können.
In mountainous Switzerland, avalanche barriers, tunnels and bridges are part of everyday architectural routine. Projects like the Sunniberg Bridge prove that engineering works for drivers can be integrated into the majestic landscape with spirit and elegance.

der weg als ziel: ting-kau-brücke, hongkong

Die Ting-Kau-Brücke verbindet die Insel Tsing Yi mit dem chinesischen Festland und ist eine der drei Brücken, die den neuen Flughafen von Hongkong (Architekt Sir Norman Foster) mit der Stadt verbinden. Es handelt sich um eine mehrfeldrige Schrägkabelbrücke, die mit ihrer 1.177 m langen Fahrbahn zu den längsten der Welt gehört. Die speziellen Bedingungen eines Taifungebiets und einer unter der Brücke kreuzenden Hauptwasserstraße hatten eine aufwendige Bauweise und komplexe Konstruktion zur Folge. So besitzt die Brücke nur drei Hauptstützen, zwei am Ufer und einen etwa 200 m hohen Mittelturm, der im Meer auf einer Unterwasser-Felskuppe gegründet wurde. Von dort aus wurden gut 500 m Fahrbahn im freikragenden Vorbau erstellt. Idee und ingenieursmäßige Umsetzung müssen im Nachhinein wegen der Tatsache, dass „man den guten Willen chinesischer Schweißer ursprünglich mit Fachwissen verwechselt hatte" (FAZ) besonders gewürdigt werden. Von einem planmäßigen und strukturierten Planungs- und Bauvorgang war deswegen kaum die Rede, der Weg, genauer die Brücke, wurde zum Ziel.

Ting-Kau-Brücke, Hongkong
Autobahnbrücke zwischen der Insel Tsing Yi und dem chinesischen Festland

Bauzeit	1995–1998
Länge	1.177 m
Konstruktion	Schrägseilbrücke
Höhe der Pylonen	max. 201 m
Ingenieure	Schlaich, Bergermann und Partner, Stuttgart

Etwa 30 km von Hongkong entfernt: die melonengelb gestrichene Ting-Kau-Brücke, die mit nur drei Türmen auskommt, um den 900 m breiten Meeresarm zum neuen Flughafen von Hongkong zu überspannen.

Etwa 30 km von Hongkong entfernt: die melonengelb gestrichene Ting-Kau-Brücke, die mit nur drei Türmen auskommt, um den 900 m breiten Meeresarm zum neuen Flughafen von Hongkong zu überspannen.
Roughly 30 kilometres from Hong Kong: with its coat of melon-yellow paint, the Tin Kau Bridge needs only three towers to bridge the 900 metre wide inlet and create a link to Hong Kong's new airport.

the goal of the way: tin kau bridge, hong kong

The Tin Kau Bridge links Tsing Yi island with the Chinese mainland, and is one of the three bridges connecting Hong Kong's new airport (architect: Sir Norman Foster) with the city. With a 1,177 metre long deck, this multi-span cable-stayed bridge is one of the longest in the world. The special conditions of an area affected by typhoons and of a major waterway passing under the bridge meant that it was complicated to design and build. It is for this reason that the bridge has only three main piers, one on each bank and one roughly 200 metre high central tower, whose foundations are sunk on a rock under the sea. From this central tower, well over 500 metres of deck were built by cantilevering out in each direction. With hindsight, the idea and the engineering feat itself deserve special praise: as the Frankfurter Allgemeine Zeitung commented, "the Chinese welders' goodwill was first of all wrongly taken to be expertise". It was there-

fore hardly appropriate to speak of a scheduled, structured process of planning and construction: the goal became one of the way (or rather of the bridge).

Tin Kau Bridge, Hong Kong
Motorway bridge between Tsing Yi island and the Chinese mainland

Constructed	1995–1998
Length	1,177 m
Structure	Cable-stayed bridge
Maximum pylon height	201 m
Engineers	Schlaich, Bergermann und Partner, Stuttgart

superharfe: erasmusbrücke, rotterdam

Rotterdam galt zur Jahrtausendwende als heimliche Architekturhauptstadt
Europas und der dort arbeitende Architekt Rem Koolhaas als der Superstar
unter den Urbanisten des 20. Jahrhunderts. Seinem Schüler, dem Architekten
Ben van Berkel (UN Studio), war es vergönnt, eines der wichtigsten Zeichen
für den Aufbruch Rotterdams in eine neue Zeit zu setzen. Van Berkel entwarf
die neue Erasmusbrücke, die den alten Kern Rotterdams mit der neuen
Erweiterung nach Süden, mit dem programmatischen Namen Kop van Zuid
verbindet. Dort sind wichtige Architekten der Welt tätig geworden für Wohn-,
Büro- und Kulturbauten (u.a. Foster, Wilson, Bolle-Wilson). Die elegante
Superharfe der Erasmusbrücke macht deutlich, dass Brücken mitten in der
Stadt nicht mehr ausschließlich Aufgabe von Ingenieuren sind. Längst haben
Architekten Brücken als wichtige Bauaufgabe für sich erkannt. Van Berkel
stellt die Brücke unter das Thema der Asymmetrie, nicht zuletzt deshalb,
um auf die Unterschiede zwischen der alten Stadt und dem neuen Stadtteil
hinzuweisen.

Bauzeit	1990–1996
Länge	802 m
Höhe	139 m
Hauptspannweite	280 m
Architekten	UN Studio van Berkel & Bos, Amsterdam

super harp: erasmus bridge, rotterdam

At the turn of the millennium, Rotterdam was regarded as the secret capital of European architecture. It is also home to the architect Rem Koolhaas, the superstar among the urbanists of the 20th century. His student Ben van Berkel (UN Studio) was granted the privilege of setting one of the most important signs for Rotterdam's entry into a new age. Van Berkel designed the new Erasmus Bridge, which links the old centre of Rotterdam with the new district being built to the south, known by the programmatic name Kop van Zuid. Major international architects (including Foster, Wilson, Bolle-Wilson) have designed residential, office and cultural buildings there, and the elegant super harp formed by the Erasmus Bridge clearly shows that city-centre bridges are no longer solely the task of engineers. Architects have long since come to realize that bridges are an important construction task for them. For the bridge, van Berkel's design theme was one of asymmetry. One reason for this, in the end, was to allude to the differences between the old city and the new district.

Constructed	1990–1996
Length	802 m
Height	139 m
Main span	280 m
Architects	UN Studio van Berkel & Bos, Amsterdam

Im Zeichen der Asymmetrie: Dazu gehört auch ein Knick im 139 m hohen Pylon, der dafür sorgt, dass die Brücke zum „raumübergreifenden und städtebaulichen Objekt" (UN Studio) wird.
Sign of asymmetry: the kink in the 139 metre high pylon, which helps to make the bridge, as UN Studio say, "a landmark and an urban development feature".

fünf tore: stadttore für san marino

Wenn man so will, ist San Marino ein Staat im Staat, vollkommen eingeschlossen durch die Republik Italien, und ist als selbstständiges Staatswesen eher durch seine Briefmarken als durch eigene Politik bekannt. Trotzdem setzt San Marino Zeichen für seine Besucher und hat sich fünf Varianten eines modernen Stadttores bauen lassen. Die Architekten des Büros Giancarlo de Carlo (mit Paolo Castiglioni, Antonio Troisi) haben eine „Stadttor-Familie" kreiert, deren Mitglieder jeweils die Grenze zwischen San Marino und Italien markieren. Die größte ist die Porta di Dogana an der Straße nach Rimini, eine gelungene Mixtur aus Landmarke, Fahnenmast und Fußgängerbrücke über die Autostraße. Alle fünf erwarten die Autobesucher wie eine lustige Camouflage aus der maritimen Welt, mit aufstrebender Mastentektonik und dem strahlenden Weiß von Schiffsaufbauten. Ein Land, das keinen Hafen, keine Küstenlinie, aber sehr, sehr engen Blickkontakt zur Adria besitzt.

Im Übrigen: Die fünf von San Marino sind ein Beweis dafür, dass durch besondere Architektur der Straßenraum an Kontur und Vielfalt gewinnt.

Baujahr	1996
Architekt	Giancarlo de Carlo, Mailand
Ingenieur	Giuseppe Carniello, Mailand

five gates: city gates for san marino

One could describe San Marino as a state within a state, completely surrounded by the republic of Italy, and famous as an independent country today more for its stamps than for its government policies. Nevertheless, San Marino sets signs for its visitors, and has had five variants of a modern city gate built by an architect. The architects from the Giancarlo de Carlo bureau (together with Paolo Castiglioni and Antonio Troisi) have created a "city gate family", whose members demarcate the frontier between San Marino and Italy. The largest of these is the Porta di Dogana on the road to Rimini, a successful mixture of landmark, flagpole and pedestrian bridge over the *autostrada*. All five of them greet motorized visitors with amusing allusions to the seafaring world, with mast structures rising into the sky and the brilliant white usually associated with ship superstructures – welcome to your cruise to San Marino. A country that has no port and no coastline, but very, very close visual contact to the Adriatic!

For the rest – San Marino's five gates are proof that outstanding architecture can add visage and variety to the road environment.

Constructed	1996
Architect	Giancarlo de Carlo, Milan
Engineer	Giuseppe Carniello, Milan

Mischung aus Tor, Brücke und Stadtzeichen: Porta die Dogana, San Marino
A pastiche of gate, bridge and urban landmark: Porta di Dogana, San Marino

47

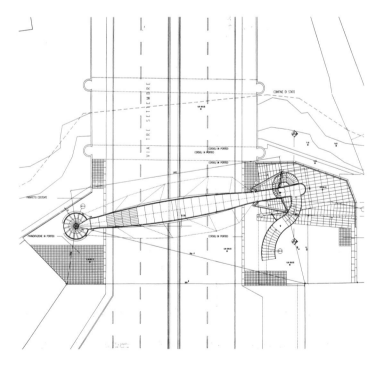
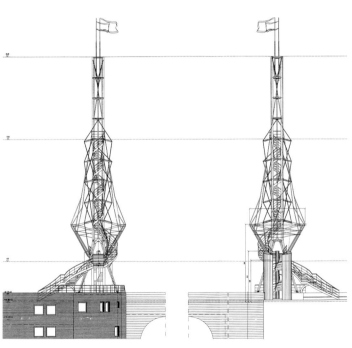

Details und Zeichnungen der Porta di Dogana

Details and drawings of Porta di Dogana

endlos, ortlos, unsichtbar:
autobahnmeisterei, nanterre

Von der Eisenbahn kennt man die Entwicklung, Bahnlinien und Bahnhöfe in den dicht bebauten Innenstadtdistrikten zu über- oder unterbauen. Das bekannteste Beispiel in Deutschland ist die Berliner Stadtbahn, die als Hochbahn durch Charlottenburg, Tiergarten und Mitte führt und deren Stadtbahnbögen Heimat für Läden, Restaurants oder Werkstätten geworden sind. Ähnlich war das Motiv in Nanterre bei Paris, wo einer eleganten Autobahnbrücke Facilities für den Autobahnbetreiber „untergeschoben" wurden. Es ist der gelungene Versuch, Bauten, die einen Bezug zur Straße haben, so unterzubringen, dass die Landschaft nicht noch weiter zugebaut werden muss. Natürlich benutzen die Architekten Decq und Cornette die erwarteten Metaphern von Mobilität und Dynamik – man könnte einen rasenden Truck darin sehen, der Phantasie sind keine Grenzen gesetzt. Aber keine Sorge: Es ist nur eine Immobilie, die den Eindruck von Bewegung erzeugt.

Die Organisation ist einfach und logisch. Empfangshalle, Büros und ein Kontrollzentrum hängen als Hauptbaukörper direkt unter den Fahrbahnen und einem Brückenträger. Vom unteren Parkplatz aus gesehen scheinen sie zu schweben. Unter und neben der Station entstand ein gedeckter Freiraum für Park- und Lagerflächen. Für den Autofahrer oben bleibt alles weitgehend unsichtbar, bis auf den Sendemast als feine Nadel auf dem Mittelstreifen der Fahrbahnen. In Nanterre ist auf diese Weise ein völlig neuer Ausdruck für eine MotorTecture entstanden, von der man nicht weiß, ob sie eine Autobahnbrücke oder ein Transitgebäude ist oder gar eine Art Flugzeugträger für Autos. In jedem Fall Architektur fürs Auto.

Bauzeit	1994–1996
Länge	250 m
Feldweiten	ca. 7 x 35 m
Konstruktion	Balkenbrücke
Architekten	Benoît Cornette, Odile Decq, Paris
Tragwerksplanung	Ove Arup & Partners, London, RFR Consulting Engineers, Paris

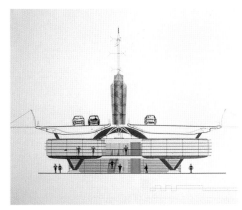

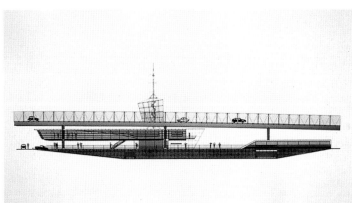

without time or place, invisible:
motorway maintenance unit, nanterre

Building railways and stations above or under ground is a solution that is well known in densely populated inner-city areas. The best known example in Germany is Berlin's commuter railway, which runs above ground through Charlottenburg, Tiergarten and the central district, and whose railway arches have become home to shops, restaurants or workshops. In Nanterre near Paris, the motive was a similar one, housing motorway maintenance facilities under an elegant motorway bridge. It is a successful attempt to situate buildings relating to the road in such a way that no further land needs to be concreted over. Of course, the architects Decq and Cornette used the usual metaphors of mobility and dynamism — it might be a snapshot of a speeding lorry, the beholder's imagination can run wild. But don't worry: it is only a building which creates the impression of movement.

Its organization is simple and logical. The reception area, offices and a control centre are suspended as the main building unit directly beneath the road decks and a bridge support. Seen from the car park below, they appear to float.

Below and next to the operations centre, there arose a covered area for parking spaces and storage. For the driver on the motorway, all of this is largely invisible, were it not for a delicate transmission mast, like a fine needle on the central reservation. In this way, a completely new expression for MotorTecture has been created in Nanterre: one where the beholder is unsure whether it is a motorway bridge or a transit building. Or, indeed, a kind of aircraft carrier for cars. At all events, this is architecture for the car.

Constructed	1994–1996
Length	250 m
Span widths	approx. 7 x 35 m
Structure	Beam bridge
Architects	Benoît Cornette, Odile Decq, Paris
Planning of supporting structure	Ove Arup & Partners, London, RFR Consulting Engineers, Paris

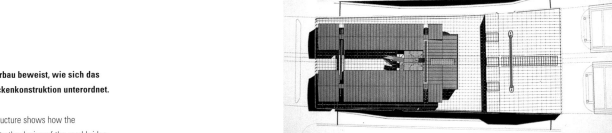

frei und filigran: tankstelle *elsa*, putbus

Weil übersteigertes Wirtschaftlichkeitsdenken und eine entsprechende Umsatzerwartung den stetig wachsenden Flächenbedarf pro Tankstellenein- heit mit sich gebracht haben, ist die Tankstelle in der Folge von einem städtischen zum vorstädtischen Bautypus mutiert. Ihr Visavis sind heute die Protagonisten des Urban Plankton: Gewerbebauten, Hamburgerbratereien und Factory Outlets. Die Tankstelle *Elsa* liegt auch am Stadtrand, allerdings am Rand der klassizistischen Residenzstadt Putbus auf Rügen und direkt am Bahnhof. Sie beweist mit ihrer individuellen und filigranen Gestalt, dass es auch anders geht, als die großen Benzinkonzerne es suggerieren. *Elsa* ist „frei", sie braucht sich also weder grün, blau oder gelb-rot in Konzernfarben zu verkleiden, sondern wählt für alle Teile, die gestrichen werden müssen, tatsächlich klassisches Weiß. Der Rest, wie die Aluminium- Dachprofile, glänzt matt silbern und metallisch. >

free and filigree: *elsa* filling station, putbus

Because filling stations need ever more space, concerns about cost-effectiveness and sales expectations have become exaggerated, and the result has been that filling stations have been transformed from a city-centre to a city-outskirt pheno- menon. Their neighbours there are the modern protagonists of urban plankton: industrial buildings, fast-food joints and factory outlets. The *Elsa* filling station is also on the outskirts of town: in this case, it is right next to the station on the outskirts of Putbus, a Classicist royal seat on the island of Rügen. Its indivi- dualistic, filigree form is proof that filling stations can be different from the standards set by the large petrol multinationals. *Elsa* is "free", i.e. independent of any one supplier, and so does not need to dress up in green, blue or yellow-red livery, but has in fact chosen classical white for all those parts that need to be painted. The rest of the building, such as the aluminium roof sections, have a matt-silver, metallic sheen. >

Wirkt verführerisch, nicht wie eine gewöhnliche Tankstelle, und wird so zum Rettungsversuch für einen malträtierten Bautypus: Tankstelle *Elsa* in Putbus auf Rügen.

Seductive in effect and unlike the run-of-the-mill filling station, it is an attempt to redeem a maltreated building type: *Elsa* filling station in Putbus on Rügen.

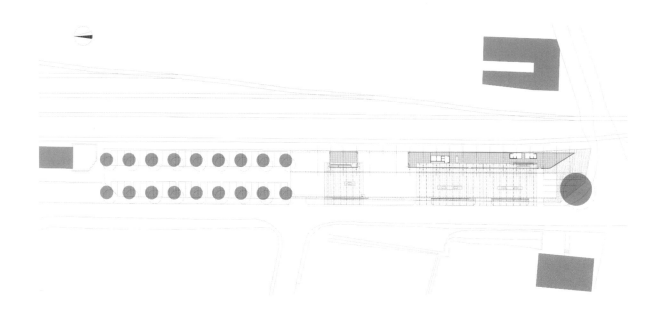

Das Grundstück ist 160 m lang, nur 18 m breit und grenzt an Bahngleise. Die Bauteile wurden parallel geparkt wie zwei Züge: Rechts am Bahnkörper liegen hintereinander ein Bistro, Kasse, Laden und Werkstatt. Die gläserne Spitze des Bistros (die Tankstelle als Treffpunkt!) und das halbrunde Tonnendach des Werkstattflügels kokettieren mit so manchem Tankstellendetail der 50er-Jahre: halbrund gebogene Blechpaneele für die Bedachung, hier wird nichts verkleidet und nichts kaschiert. Eine gewisse Rohbauästhetik und Materialienwelt, wie sie aus Werkstätten seit eh und je bekannt ist, bleibt erhalten. Als transparente Glasschatulle präsentiert sich im hinteren Baukörper die Waschanlage, die alle Geheimnisse der Waschautomatik preisgibt und Technik wie auf einem Altar präsentiert.

Das Fazit solcher Strukturen: Sie sind leicht, luftig und improvisiert, vor allem ohne Anspruch auf Ewigkeit. Dieser wird wenige Meter oberhalb in der historischen Idealstadt genügend erhoben und auch eingelöst. Hier unten wirkt alles eher durchschaubar, flüchtig und für den Augenblick gemacht. Vielleicht fliegt das Dach, das an Tragflächen von Großraumjets erinnert, tatsächlich irgendwann davon?

Bauzeit	1997–1998
Konstruktion	Stahl/Glas
Architekt	Niclas Dünnebacke, Stralsund, Xavière Bouyer, Paris
Ingenieur	Jan Müller, Bergen

The plot on which it is built is 160 metres long and only 18 metres wide, and is bordered by railway sidings. The building elements have been set down, as it were, like two trains standing next to each other: To the right, nearest to the railway, a bistro, the cash desk, a shop and a workshop are arranged one behind each other. The bistro's glass "bow" (filling station as meeting point!) and semi-circular arched roof of the workshop wing are a teasing allusion to many of the features of 1950s filling stations: sheet-metal panels bent into semi-circles are used for the roof, and nothing here is disguised or hidden. What remains is a certain aesthetic sense of the building shell and of building materials, something that has long been familiar in workshops. In the rear building element, the car wash is like a transparent glass case, revealing all the secret workings of the car wash and presenting its technology as though it were on an altar.

The bottom line is that these structures are light, airy and improvised, and above all have no claim to eternal validity. A few metres further on, in the historic city, this claim is made and redeemed quite well enough. Down here at the filling station, on the other hand, everything gives the impression of simplicity, brevity and transience. Who knows, perhaps the roof, reminiscent of a jet aircraft's wing, really will fly away one day.

Constructed	1997–1998
Structure	Steel/glass
Architect	Niclas Dünnebacke, Stralsund, Xavière Bouyer, Paris
Engineer	Jan Müller, Bergen

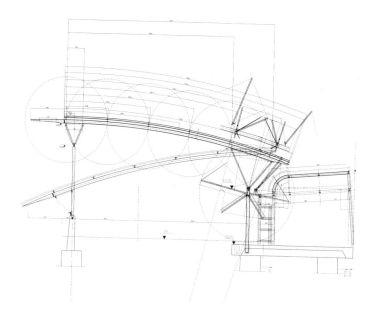

Oben: Grundriss. Rechts: Dachkonstruktionsdetail
Above: plan. Right: detail of roof construction

Tankstelle neuer Generation: Für ein 24-Stunden-Dienstleistungszentrum reicht es nicht aus, nur Benzin zu verkaufen. Auch im Angebot: unter anderem ein Bistro.

A new generation of filling station: as a 24-hour service centre, it is not sufficient just to sell petrol. The services on offer also include a bistro.

**Der Umsteigebahnhof ist einfach nur ein übergroßes Dach:
Schutz für Menschen in Bewegung und beim Warten.**

The interchange station is simply an oversize roof: protection
for people on the move or waiting.

unter bewegten dächern:
tram- und umsteigebahnhof/park and ride station, straßburg

Europas Hauptstadt Straßburg setzt nicht auf das Auto, sondern auf die Tram, wird aber auf Automobilität weder verzichten, noch sie bekämpfen. Die Stadt Straßburg macht deswegen ein Angebot, gestaltet den Verknüpfungspunkt von Eisen- und Straßenbahn mit dem Auto besonders kommod. Über ein Park-and-Ride-System wird das Auto mit dem vorbildlichen Straßburger System des öffentlichen Nahverkehrs verbunden. Die erfolgreiche irakisch-britische Architektin Zaha Hadid hatte bereits vor Jahren die Gelegenheit, mit einer dynamischen Betonskulptur für die Werksfeuerwache von Vitra in Weil am Rhein ihr Konzept von MotorTecture zu formulieren. Hier in Straßburg folgte die Fortsetzung, diesmal heißt das Thema „Bewegung", also Umsteigen, Verknüpfen und Verweben. Und so ist dann auch ihr Bauwerk zu interpretieren: als Komposition aus bewegten Teilbaukörpern, mit schrägen Stützen und fliehenden Dächern, aus immobilem Beton. Zaha Hadid hat dem einfachen, aber wichtigen Vorgang des Umsteigens ein dynamisches Gesicht gegeben.

Bauzeit _____ 1999–2001
Architekt _____ Zaha Hadid Architects, London

under dynamic roofs:
tram station and interchange/park and ride station, strasbourg

Europe's capital Strasbourg is backing the tram rather than the car, but it will neither do without nor declare war on the car. Strasbourg is therefore making an offer to drivers, and has designed the nodes where roads meet railways and tramlines to make them especially convenient. Via a park and ride system, the car is integrated in Strasbourg's excellent public transport system. Many years ago, the successful British-Iraqi architect Zaha Hadid was given the chance to formulate her MotorTecture, and created a dynamic concrete sculpture for Vitra's works fire station in Weil am Rhein. She continues this work here in Strasbourg, this time with "movement" as her theme, i.e. changing trains, linking and inter-weaving. And this is how her building can be interpreted: as a composition of dynamic building components, of leaning supports and slanting roofs and of immobile concrete, Zaha Hadid has given a dynamic quality to the simple yet important process of changing trains.

Constructed _____ 1999–2001
Architect _____ Zaha Hadid Architects, London

Seit dem Bau der Golden Gate Bridge in San Francisco (1933–37) habe es bezüglich der Konstruktion von Großbrücken nichts Neues gegeben. Teilen Sie diese Meinung eines ihrer Kollegen?

Das stimmt wahrscheinlich, soweit es die ganz großen Entwicklungsschübe betrifft. Doch hat es natürlich permanent kleine Fortschritte gegeben, zum Beispiel mit den Schrägseilbrücken für den mittleren Spannweitenbereich von 300 bis 800 Metern. Der Nachteil von Hängebrücken sind ihre gigantischen Widerlager, die teuer sind und Probleme mit der Gründung mit sich bringen. Die Schrägseilbrücke braucht das nicht, weil sie in sich selbst verankert wird und sie im freien Vorbau errichtet werden kann. Insofern ist sie der Hängebrücke überlegen.

Sie sind Ingenieur, Sie konstruieren und Sie gestalten?

Natürlich ist jeder entwerfende Ingenieur auch Gestalter! Dieser alte Konflikt zwischen Architekt und Ingenieur passt nicht in unsere komplexen Zeiten. Allerdings wird ein Ingenieur niemals aus High Tech so etwas wie High Effect werden lassen. Ich versuche einfach nur, den Kraftfluss nachzuzeichnen und in der Konstruktion zu visualisieren, wobei ich natürlich das Recht habe, mehr als das Notwendige zu machen, ich kann spielen! Zum Beispiel hätte bei der Klappbrücke in Kiel (S. 36ff) der Mechanismus auch mit einem einzigen Gelenk funktioniert. Wenn wir mehr Gelenke einführen, ist das ein Spiel mit der Kinematik, um den Bewegungsvorgang zu zeigen und genießen zu können!

Halten Sie sich damit an die alte Ingenieursaufgabe, Technik verständlich zu machen?

Gewissermaßen schon. Wir wollten mit dieser Klappbrücke zeigen, wie sie funktioniert, haben sozusagen statt des ICE-Triebwagens eine Dampflok auf High-Tech-Level gebaut. Ich meine damit, dass man bei der Dampflok ja auch von außen noch genau sehen kann, wie die Kraft über die Gestänge auf die Räder übertragen wird.

Haben Sie außer Brücken noch anderes gebaut, was mit dem Auto in einem engen Zusammenhang steht?

Ja, in München eine Wartungshalle für Fahrzeuge der Stadtreinigung und jetzt gerade einen glasüberdachten zentralen Omnibusbahnhof in der Hamburger City. Das sind interessante Ingenieursaufgaben, weil wir jeweils marktplatzgroße Überdachungen konstruieren mussten.

Wie gehen Sie dabei vor?

Wir arbeiten unspektakulär; gerade wenn man sparsam und effizient arbeiten soll, was bei „Verkehrsbauten" sozusagen immer vorausgesetzt wird, ist das dem Ingenieur angemessen. Natürlich nähern wir uns solchen Bauaufgaben anders, als man an ein Bankgebäude herangehen würde. Aber an technische Bauwerke ausschließlich mit technischem Denken heranzugehen ist nicht unbedingt zwingend!

→ **interview** interview

Jörg Schlaich ist Chef des Stuttgarter Büros Schlaich, Bergermann und Partner und zählt zu den wichtigsten deutschen Ingenieuren. Natürlich ist er ein Spezialist für Brückenkonstruktionen, hat aber auch die neue Skodafabrik (S. 112ff) und den neuen Hamburger Busbahnhof konstruktiv-kreativ begleitet und sich damit auch intensiv dem Thema Auto und Architektur gewidmet.

Jörg Schlaich is the head of the Stuttgart-based Schlaich, Bergermann und Partner bureau, and is regarded as one of Germany's foremost engineers. Of course, he is a specialist in the area of bridge construction, but was also involved in the construction of and ideas behind the new Skoda factory (p. 112ff) and Hamburg's new bus station, and has thus also devoted a lot of thought to the topic of cars and architecture.

One of your colleagues has said that there has been nothing new in the construction of large bridges since the Golden Gate Bridge was built (1933–37). Do you agree with him?

He's probably right as far as major developmental shifts are concerned. But of course, there have constantly been small steps forward. In the area of cable-stayed bridges, for example, the average span has increased from 300 to 800 metres. The disadvantage of suspension bridges is that they need huge abutments. These are expensive, and foundations for them are a problem. The cable-stayed bridge has no need of this because it is self-anchoring and can be erected by cantilevering out. In this respect, it is superior to the suspension bridge.

You are an engineer, you construct and you design?

Any engineer who designs is of course also a designer! The old conflict between architect and engineer is no longer appropriate for our complex age. Having said that, an engineer will never allow "high tech" to degenerate into "high effect". All I try to do is to follow the flow of forces and to visualize them in my design, although of course I have the right to do more than necessary and to play around a bit. In the case of the bascule bridge in Kiel, for example (p. 36ff), a single-joint mechanism would also have worked. If we bring in more joints, then we are playing with kinematics, so that we can show and enjoy the process of movement.

Are you guided by the old engineering task of making technology understandable?

Yes, to a certain extent. In this bascule bridge, we wanted to show how it works. It was as though we were building a high-tech steam locomotive rather than the drive unit of a bullet train. What I mean is that with a steam locomotive you can see precisely from the outside how forces are transmitted to the wheels via the piston rods.

Apart from bridges, have you ever built anything else closely connected with the car?

Yes, I've built a maintenance centre in Munich for municipal street-cleaning vehicles, and at the moment I'm building a central bus station in the centre of Hamburg, which will be covered with a glass roof. They are interesting engineering challenges, because in both cases we have had to design roofs to cover an area as large as a market place.

How do you go about that?

We work modestly; engineers don't mind having to work economically and efficiently, which almost always seems to be required when "buildings for traffic" are involved. I find that this is an appropriate method for engineers. Of course, we approach design tasks such as these differently from the way we would tackle a building for a bank. But there is no law which says that a technical frame of mind is the only way to approach technical buildings!

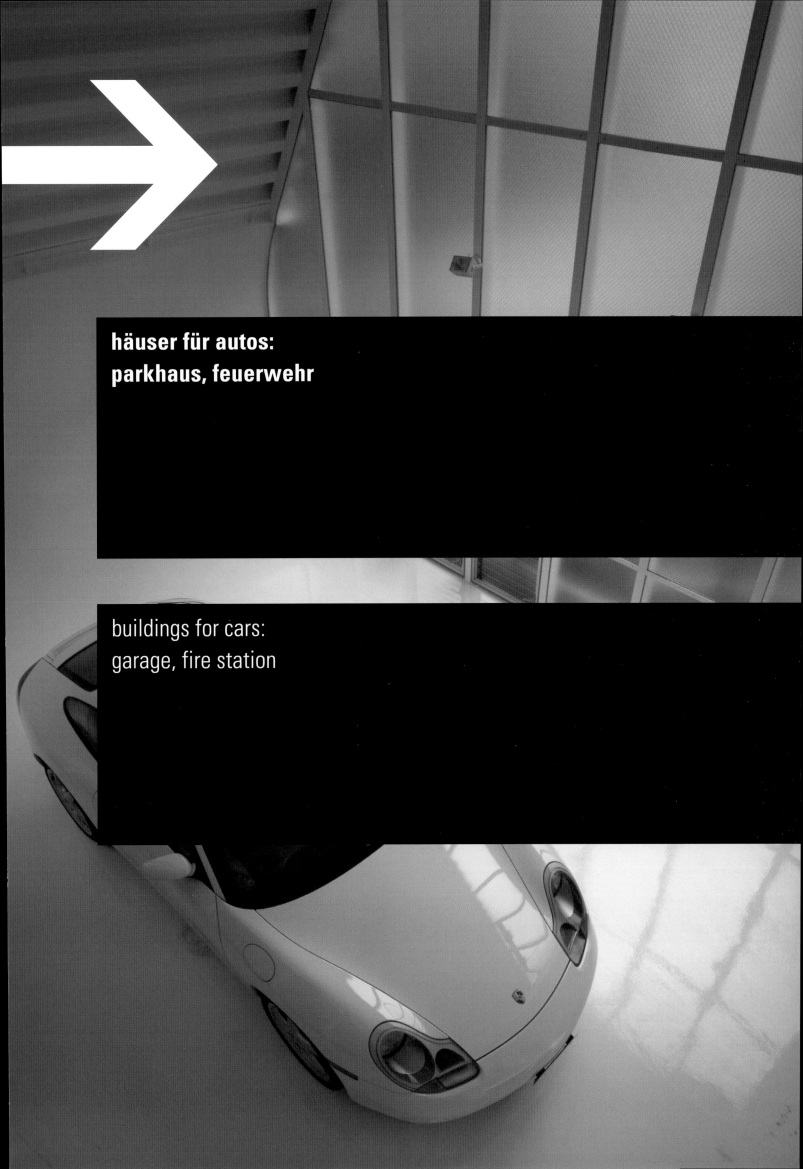

häuser für autos:
parkhaus, feuerwehr

buildings for cars:
garage, fire station

vroom! garage und showroom, nagoya

Keine Überraschung: Japans Megastädte sind zur autogeprägten und -beherrschten Stadtlandschaft geworden. Allerdings bestehen auch das moderne Tokio und die anderen Metropolen immer noch aus einer Phalanx aus wunderbar kleinteiligen Stadtvierteln mit Sträßchen, die kaum oder gar nicht für den Autoverkehr geeignet sind. Nach dem Big Bang after the Boom ist in Japan der architektonische Alltag eingekehrt, soll heißen, auch kleine Aufgaben reizen die großen Architekten, insbesondere dann, wenn sie wie Astrid Klein und Mark Dytham (KDa) aus westlichen Kulturen kommen. *Vroom* ist die Garage eines Sammlers von Luxuskarossen: ein Häuschen für Autos, integriert in die idyllische Harmlosigkeit der unendlichen Vorstadt japanischer Megastädte. Im geschlossenen Zustand erweckt das Garagentor, banal wie es dann wirkt, den Eindruck, dahinter sei nun absolut nichts zu holen. Stahlblech, Beton, Rohbauambiente – allerdings vom Feinsten. >

vroom! garage and showroom, nagoya

Not surprisingly, Japan's mega-cities have become car-scaled, car-dominated cityscapes. Even so, present-day Tokyo and the other metropolises are still made up of a phalanx of wonderfully intricate urban districts, with tiny roads that are scarcely or not at all suitable for car traffic. Following the big bang after the boom, architectural normality has returned to Japan, which means that minor tasks are also intriguing for major architects, particularly if they come from Western cultures, like Astrid Klein and Mark Dytham (KDa). *Vroom* is the garage built for a collector of luxury limousines: a little house for cars, integrated in the idyllic harmlessness of the never-ending suburbs of Japanese mega-cities. When it is closed, the banal effect created by the garage door makes sure that it gives the impression that there is absolutely nothing of interest on the other side. Sheet steel, concrete, the ambience of the building carcass – albeit with the very best materials. >

Blechkiste, Schatzkiste? Kaum ein Passant ahnt, dass dahinter Luxusautos parken.

Metal box, treasure chest? Hardly any passer-by has an inkling that luxury limousines are parked behind this door.

Hinter dem Tor kommen die Architekten allerdings richtig in Schwung, und das teilt sich im Fußboden und der entsprechend gekurvten Seitenansicht mit, die auch von Porsche- oder Maserati-Konstrukteuren entworfen sein könnten. Interessant dabei: KDa sind eigentlich Designer, die an ihren kleinen Aufgaben wie Showrooms oder Büros einen notorischen Spieltrieb entwickelt haben, sie schöpfen sonst immer aus dem Vollen (wie mit dem Kunstpflanzenensemble für die Schaufensterdekoration eines vegetarischen Restaurants in Tokio). Hier, gerade hier, nehmen sie sich seltsamerweise sehr stark zurück. Die Lösung: Das Luxusauto ist der Star, es steht extra vor hellem Uni-Hintergrund, gerade so wie moderne Grafik besonders gut an weißen Wänden zur Wirkung kommt. Zum optimalen Ambiente trägt allerdings der gekurvte Fußboden bei, es gibt keine Wand, keinen Boden, keine Decke, sondern nur einen übergangslosen Bogen vergleichbar der Fahrbahndecke einer Bergstraße: So, als wollten die geparkten Fahrzeuge gleich durchstarten …

Baujahr	1999
Architekten	Klein Dytham Architecture (KDa), Tokio

Once they get beyond this door, however, the architects really get going, as is shown by the flooring and the correspondingly curved side elevation, which could also have been designed by Porsche or Maserati designers. What is interesting here is that KDa are in fact designers who have developed a notorious playfulness in their minor assignments such as showrooms or offices, and their designs have always had a no-holds-barred quality (such as the arrangement of plastic flowers for the window decoration of a Tokyo vegetarian restaurant). Strangely, they have been extremely reserved here. Their solution: the luxury limousine is the star, placed specially before a monochrome, bright background, in the same way as modern graphic art is particularly effective against white walls. The curved floor also plays its part in creating the optimum setting – there is no wall, no floor, no ceiling, just a smooth curve like the lane of a mountain road: it is as though the parked cars were just waiting to zoom off …

Constructed	1999
Architects	Klein Dytham Architecture (KDa), Tokyo

Ein gekurvtes Passepartout für die Autobaukunst – *Vroom* in Japan.
A curved passe-partout for the carmaker's art – *Vroom* in Japan.

Außen und innen: im Zeichen der Kurve und so einfach wie ein Rohbau.
Outside and inside: dedicated to the curve, and as simple as a building carcass.

der verkehrsbau als stadtzeichen:
parkhaus am bollwerk, heilbronn

Wieder ist der Name Programm: Das Heilbronner Parkhaus am Bollwerk, also in Nachbarschaft zur ehemaligen Stadtmauer, übernimmt an historischer Stelle die „Formulierung eines Stadteinganges" (die Architekten). Die ansprechende Form des Parkhauses – ein sehr langer Riegel mit zwei abgerundeten Enden, von denen eines die notwendige Auffahrtsspindel aufnimmt – und die dezidiert gestaltete Fassade – eine zweischalige Konstruktion aus Holz und vorgehängtem Drahtgeflecht, die dem Parkhaus die übliche Härte und Belanglosigkeit nimmt – erreichen zusammen das, was Parkhäusern meist vorenthalten bleibt: Es entsteht nicht nur ein dienender Verkehrsbau, sondern ein Gebäude, das in der Architektur von Heilbronn eine sympathische Rolle als Orientierungspunkt und vielleicht bald sogar als Wahrzeichen übernimmt.

Bauzeit	1997–1998
Nutzfläche	ca. 13.500 qm, 500 Stellplätze
Konstruktion	Stahlbeton und Stahl
Fassade	Douglasien- und Lärchenprofil-Holzlamellen und vorgelagertes verzinktes Drahtgeflecht
Architekten	Mahler, Günster, Fuchs, Stuttgart
Statik	Fischer + Friedrich, Stuttgart

a building for cars as an urban landmark:
bollwerk multi-storey car park, heilbronn

Again, the name is a manifesto: Heilbronn's Bollwerk (bastion) multi-storey car park, as the name suggests, is close to the former town walls, and in this historic context assumes, in the architects' words, the task of "formulating the entrance to the town". The bold form of the multi-storey car park – an extended oblong with two rounded ends, one of which houses the obligatory spiral ramp – and its unashamedly designed façade – a double-shell construction of wood and bracket-mounted wire meshing, which takes away the severity and triviality common to most car parks – combine to achieve an effect that is lacking in the average multi-storey car park. This is not just a building to serve the needs of traffic, but an appealing building in the ensemble of architectural styles in Heilbronn, which has assumed the role of a landmark and may even soon become an emblem.

Constructed	1997–1998
Usable area	approx. 13,500 m², 500 parking spaces
Structure	Reinforced concrete and steel
Façade	Strips of Douglas fir and larch, in front of which galvanized wire meshing has been suspended
Architects	Mahler, Günster, Fuchs, Stuttgart
Structural engineer	Fischer + Friedrich, Stuttgart

Wie ein Schleier liegt über der rauen Holzverschalung ein Maschendrahtgeflecht:
aus einem simplen Verkehrsbau wird ein sympathischer Bestandteil der Stadtumwelt.
Wire meshing shrouds the rough wood boarding like a veil: a simple building for cars
is transformed into an attractive feature of the town environment.

maschine, safe, regal:
parkregal posenerstraße, sindelfingen

Nicht verwunderlich: In der Autoregion Stuttgart wächst der Bedarf an Parkraum. In Sindelfingen besann man sich auf Ideen der Auto-Urzeit und etablierte ein einfach strukturiertes, platzsparendes Parkregal. Dahinter verbirgt sich schlicht eine „Parkmaschine" mit Aufzügen, die auf engstem Raum, das heißt auf gut 300 qm, 124 Autos wegstapeln kann. Damit wird ein herkömmlicher Parkplatz von 2.565 qm ersetzt. Vier Aufzüge und Parkpaletten bilden mit offen liegender Konstruktion diese Maschine und versorgen neun Parkdecks. Eine Glasfassade aus punktgehaltenen Glasscheiben schützt die Autos vor Wind und Wetter und erzeugt ganz nebenbei eine Anmutung von Wert und Sicherheit: „Parksafe" nennen die Architekten ihre Konstruktion auch gern.

machine, safe, shelf:
posener strasse multi-storey car park, sindelfingen

It is no surprise that demand for parking space is growing in the car-manufacturing region around Stuttgart. Sindelfingen has rediscovered ideas from the first days of the car and set up a simply structured, space-saving parking silo. When it comes down to it, this is a simple "parking machine" with lifts, which can store away 124 cars while taking up a minimum of space (just over 300 square metres). This is instead of a conventional car park covering 2,565 m². Four lifts and parking pallets make up this machine, whose construction is open for all to see, and serve nine parking decks. A glass façade made up of pointwise supported glass panes protects cars from the elements, and creates as a side-effect a feeling of value and security: indeed, the architects like to call their structure a "park safe".

„Parksafe": Eine täglich wechselnde Auto-Show, weil die PKW täglich neu geparkt werden, sorgt für ein sich stetig wandelndes Haus-Bild.
"Park safe": Every day a new motor show: with cars being parked in different places every day, the result is a building whose appearance is constantly changing.

Bauzeit	1997–1998
Nutzfläche	331 qm, 46,8 m x 6,65 m, auf sieben oberirdischen und zwei unterirdischen Ebenen, 124 Stellplätze
Konstruktion	Stahlbeton und Stahl
Fassade	2,93 x 1,80 m Glastafeln, punktgehalten, Stahlkonstruktion
Architekten	Petry + Wittfoht, Stuttgart, Frankfurt
Statik	Fischer + Friedrich, Stuttgart

Constructed	1997–1998
Usable area	331 m², 46.8 x 6.65 m, on seven overground and two underground levels, 124 parking spaces
Structure	Reinforced concrete and steel
Façade	Glass panes 2.93 x 1.80 m, supported pointwise, steel structure
Architects	Petry + Wittfoht, Stuttgart, Frankfurt
Structural engineer	Fischer + Friedrich, Stuttgart

Autos hinter Glas: Ausdruck von Wert und Sicherheit

Cars behind glass: impression of value and security

Das Parkregal funktioniert wie eine Maschine. Aufzüge und Parkpaletten sorgen für Flexibilität und Bewegung auf kleinstem Raum. Unten: Übersicht und Schnitte
The multi-storey car park is like a machine. Lifts and parking pallets allow flexibility and manoeuvrability in a minimum of space. Below: Outline and section

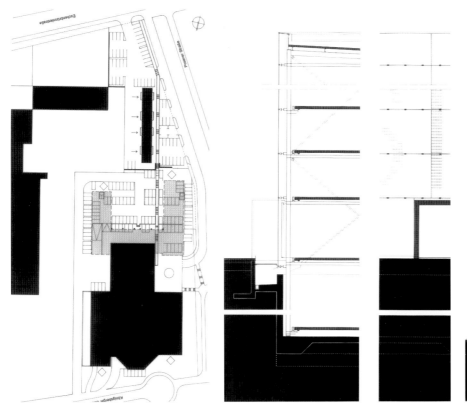

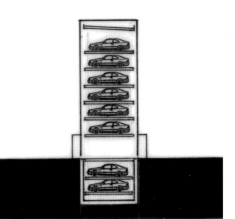

ein reptil, das nie schläft: feuerwache, maastricht

Zum schier unversiegbaren Quell niederländischer Architektenkraft gehört seit gut zehn Jahren auch das Duo Neutelings und Riedijk aus Rotterdam. Ihre Bauten zählen dabei zum Überraschendsten und Humorvollsten, was die an sich kalvinistisch geprägten Holländer in den letzten Jahren hervorgebracht haben. Das ist wenig verwunderlich, wenn man weiß, dass Neutelings und Riedijk die Kombination aus Faulheit und Ehrgeiz in der Architektur schätzen. Statt diese Methode nun weiter zu analysieren, genügt der Hinweis darauf, dass ihre Bauwerke sich meist durch ein Augenzwinkern auszeichnen. Willem Jan Neutelings und Michiel Riedijk haben sich gleich mehrfach dem Thema Feuerwehrhaus gewidmet, einer Bauaufgabe, um die sich auch die Großen der Branche wie Robert Venturi oder Zaha Hadid kümmerten. Warum? Vielleicht, weil der früher obligatorische Turm zum Trocknen der Wasserschläuche einer Feuerwehrstation die Sakralität einer Kirche oder mindestens einen Anschein davon verleihen konnte. >

Bauzeit _____ 1996–1999
Architekten _____ Neutelings Riedijk, Rotterdam

a reptile that never sleeps: maastricht fire station

For a good ten years now, the Rotterdam-based duo Neutelings and Riedijk have been an almost inexhaustible source of architectural vitality. Their buildings are some of the most surprising and witty structures to have been produced by the Dutch – a Calvinist nation, after all – in recent years. This is less of a surprise when one realizes that their creed speaks of a combination of laziness and ambition in architecture. But instead of analyzing this method further, it will suffice to say that their buildings are usually typified by a sense of playful irony. Willem Jan Neutelings and Michiel Riedijk have devoted their attention several times to the fire station theme, an architectural task that has also occupied some of the great names in architecture, such as Robert Venturi or Zaha Hadid. Why? Perhaps because the fire station tower that used to be obligatory to dry hose pipes might bestow the mystic quality of a church, or at least a pastiche of it. >

Constructed _____ 1996–1999
Architects _____ Neutelings Riedijk, Rotterdam

Die Architekten nehmen die Grundidee ihres Entwurfs mit einem Comic auf: das schlafende Reptil auf dem Sprung?
The architects use a comic to record the basic idea behind their draft design: a sleeping reptile waiting to pounce?

Liegt zentral und immer einsatzbereit am nördlichen Autobahnring von Maastricht: die neue Feuerwache von Neutelings und Riedijk.

In a central location, always ready for action on Maastricht's northern motorway by-pass: the new fire station by Neutelings and Riedijk.

Die Brandweer in Maastricht verbreitet nichts von dieser Sakralität. Sie wirkt zunächst ganz trivial und ist erst einmal das, was ein Feuerwehrhaus sein sollte: die Kombination aus Büro, Werkstatt und Garage für Spezialfahrzeuge. Geschickt werden die Schlafräume der Alarmbereitschaft im hinteren Obergeschoss neben einen Dachgarten gelegt. Dieser Dachgarten versorgt die unten liegende Halle auch in der Tiefe des Raumes mit Oberlicht.

Mit dem obligatorischen „Comic" zeigen die Architekten ihren Sinn für Humor. Dort bilden sie ihr Feuerwehrhaus wie ein scheinbar schlafendes Reptil ab. Denn durch den Trick, die Fahrzeughalle in der Front einzuziehen und das Obergeschoss auskragen zu lassen, entsteht der flüchtige Eindruck eines großen Mauls, das bei Gefahr viele Feuerwehrautos ausspucken wird. Das Haus besitzt zwar keine Reptilienhaut, geht aber mit einer „Schuppung" in der Fassade durchaus in diese Richtung und nimmt geschickt den Gedanken einer „Auto-Architektur" auf: die Muster sind nach zahlreichen Studien von Reifenprofilen entstanden.

The "Brandweer" in Maastricht does not give this impression at all. At first sight, the impression it gives is quite trivial, and the building is everything a fire station should be: a combination of office, workshop and garage for special-purpose vehicles. The dormitories for the crews on alert have conveniently been placed at the rear of the first floor, next to a roof garden. This roof garden provides light from above, reaching into every corner of the hall below.

The playful irony starts with the "comic", a hallmark of these architects' work. They have painted the fire station to look like a sleeping reptile. By using the trick of making the ground-floor vehicle hall recede into the façade and allowing the first floor to project, the fleeting impression is created of a huge mouth that will spit out lots of fire engines when danger approaches. While the building does not have the horny skin typical of reptiles, the "scales" on its façade are by all means a hint in this direction, skilfully taking up the idea of "car architecture": the patterns were the result of several studies of car-tyre treads.

Das kompakte Gebäude ist klar gegliedert: vorn die Fahrzeughalle, dahinter Werkstätten und Facilities. Oben: Schlafräume, Büros und anderes.
The compact building is clearly structured: vehicle hall at the front, workshops and facilities to the rear. On the first floor: dormitories, offices and other rooms.

maulwurfshügel: straßenservicegebäude, harlingen

Harmloser kann eine Bauaufgabe nicht mehr sein: Für das niederländische Reichsamt für Straßen und Wasserwege wurde in Harlingen ein Versorgungs- und Verwaltungsgebäude gebaut, um hier vor allem Baumaterialien oder Streugut zu lagern und den entsprechenden Versorgungsfahrzeugen einen Unterschlupf zu bieten. Die Fassade besteht aus schwarzen, einfachen, gewellten Betonplatten. Der einzige architektonische Trick liegt in den durchweg schräg gestellten Fassaden. So wird aus einer üblicherweise langweiligen Kiste so etwas wie ein abstrakter Maulwurfshügel.

Bauzeit _____ 1996–1998
Architekt _____ Neutelings Riedijk, Rotterdam

molehill: supply and administrative centre, harlingen

The building task could hardly have been more harmless: a supply and administrative centre has been built in Harlingen for the Dutch highways and waterways office, as a location in which to store building materials or grit and as a shelter for the concomitant supply vehicles. The façade is made up of black, simple, corrugated concrete slabs. The only architectural trick is the oblique angle at which all the facades have been set. In this way, what would usually be a boring box has become a kind of abstract molehill.

Constructed _____ 1996–1998
Architect _____ Neutelings Riedijk, Rotterdam

Der Comic zum Gebäude.
The comic to go with the building.

Warum werden Parkhäuser und Tiefgaragen architektonisch so stiefmütterlich behandelt?

Man hat bisher Parkhäuser vorwiegend als Aufbewahrungsort für Autos begriffen und deshalb so wirtschaftlich wie möglich nach dem Motto „Hauptsache es regnet nicht rein" betrieben. Die Tiefgarage ist dabei häufig eher Nebensache. Die hochwertigen Nutzungen für die Wohnungen und Arbeitsplätze der Menschen liegen überirdisch, die Autos stehen im Keller; da steckt ja auch eine gewisse Logik dahinter. Was nun überirdische Parkhäuser betrifft, mangelt(e) es den Architekten oft an Phantasie.

Ist Besserung in Sicht?

Die Einstellung der Bauherren und Betreiber hat sich verändert, weil auch Parkhäuser heute in einem Wettbewerb um Garagenbenutzer stehen. Allerdings stehen erst einmal harte Kriterien wie Erschließung und Sicherheit im Mittelpunkt. Dafür steht das Thema Frauenparkplätze wie auch die Gestaltung der Wege vom und zum Fahrzeug. Allerdings werden heute Parkhäuser häufiger zum Gegenstand eines Architektenwettbewerbs und werden nicht mehr nur den so genannten Spezialisten überlassen. Das sagt einiges über wachsende Ansprüche an die Gestaltung aus.

Was kann man denn nun an einem Parkhaus noch weiter verbessern? Sind es Aspekte des Komforts, der Ausstattung oder sogar auch eines anderen Nutzungsmixes?

Das Beispiel eines unserer neuen Parkhausprojekte in der Innenstadt von Münster beweist: Es geht um mehreres. Dieses Parkhaus ist ein Mixed-Use-Gebäude, im oberen Geschoss ist die Verwaltung untergebracht. Wir nutzten dies für eine relativ expressive Glasaufstockung. Das Parkhaus selbst besitzt eine geschuppte, transluzente Glasfassade, die Autos bleiben sichtbar, es entsteht eine sich stets ändernde belebte Kulisse, die der Straße an dieser Stelle gut tut. Außerdem hat das Parkhaus jeweils unterschiedliche, aber ausdrucksstarke Tages- und Nachtgesichter.

Sie haben einmal gesagt, Fahrzeuge sind eigentlich Stehzeuge, weil sie meistens pro Tag nicht mehr als eine Stunde in Bewegung sind. Dann sind doch die Maßnahmen für den ruhenden Verkehr besonders notwendig und wichtig. Sollte man das nicht auch mit der Gestaltung der Parkhäuser ausdrücken?

Also Parkhäuser werden niemals wie Kathedralen aussehen, dazu wird das Thema auch weiterhin zu sehr vernachlässigt. Aber wir werden weiter an interessanten neuen Wegen arbeiten, beispielsweise die Idee der kurzen Wege im Parkhaus untersuchen und Kombi-Nutzungen anbieten: mit Parkregalen oder auch Parkraum auf den geneigten Ebenen, die gleichzeitig die Garage erschließen und den Parkhäusern individuelle Gesichter geben.

→ **interview** interview

Die Architekten Falk Petry und Jens Wittfoht führen ein Architekturbüro in Stuttgart und beschäftigen sich unter anderem verstärkt mit Verkehrsbauten. Von ihnen stammt das Parkregal Posenerstraße in Sindelfingen (S. 68 ff).
The architects Falk Petry and Jens Wittfoht run an architects' office in Stuttgart. Buildings for traffic are one of their main areas of work. They designed the parking silo in Sindelfingen's Posener Strasse (p. 68 ff).

Why are multi-storey car parks and underground car parks given such scant attention architecturally?

Up to now, car parks have generally been regarded as a place of safe-keeping for cars. That's why they have been run as economically as possible – all that matters is that the cars don't get wet. In this context, an underground garage is frequently more of an incidental item. High-status uses, such as apartments or workplaces, are situated above ground, while cars are kept in the basement: one cannot deny that there is a certain logic behind it. As far as multi-storey car parks are concerned, architects frequently lack (or have lacked) fantasy.

Are there any signs of improvement?

Clients' and operators' attitudes have changed, not least because modern garages are competing with each other for customers. First of all, however, the focus is on hard criteria such as access and security/safety. This includes topics such as parking spaces for women and the design of access routes to and from the vehicle. Having said that, multi-storey car parks are now more frequently the subject of architectural competitions, and are no longer being left solely to the "specialists". That is a sign that the demands made of design are growing.

What is there about a multi-storey car park that can be improved? Is it features such as convenience, equipment or, indeed, combining it with other uses?

The example of one of our new car park projects in the city centre of Münster shows that several aspects are involved. This multi-storey car park is a mixed-use building, its top storey housing administrative offices. We have exploited this fact and added on a relatively expressive glass element. The multi-storey car park itself has a scaly, translucent glass façade, behind which the cars remain visible. The result is a lively backdrop that is always changing, which is a welcome feature on this part of the road. Moreover, the car park has two different yet equally expressive faces, during the day and at night.

You once said that vehicles are actually stationary objects, because they are frequently not in motion for longer than one hour a day. That makes actions to accommodate stationary vehicles especially necessary and important. Shouldn't this also be expressed in car park design?

Car parks will never look like cathedrals – the topic is still far too neglected for that to happen. But we will continue to work on interesting new ideas; looking for example at the idea of short walking distances in the car park and offering combined uses: with parking silos or even parking space on the sloping levels that also provide access to the garage and give multi-storey car parks their individual appearance.

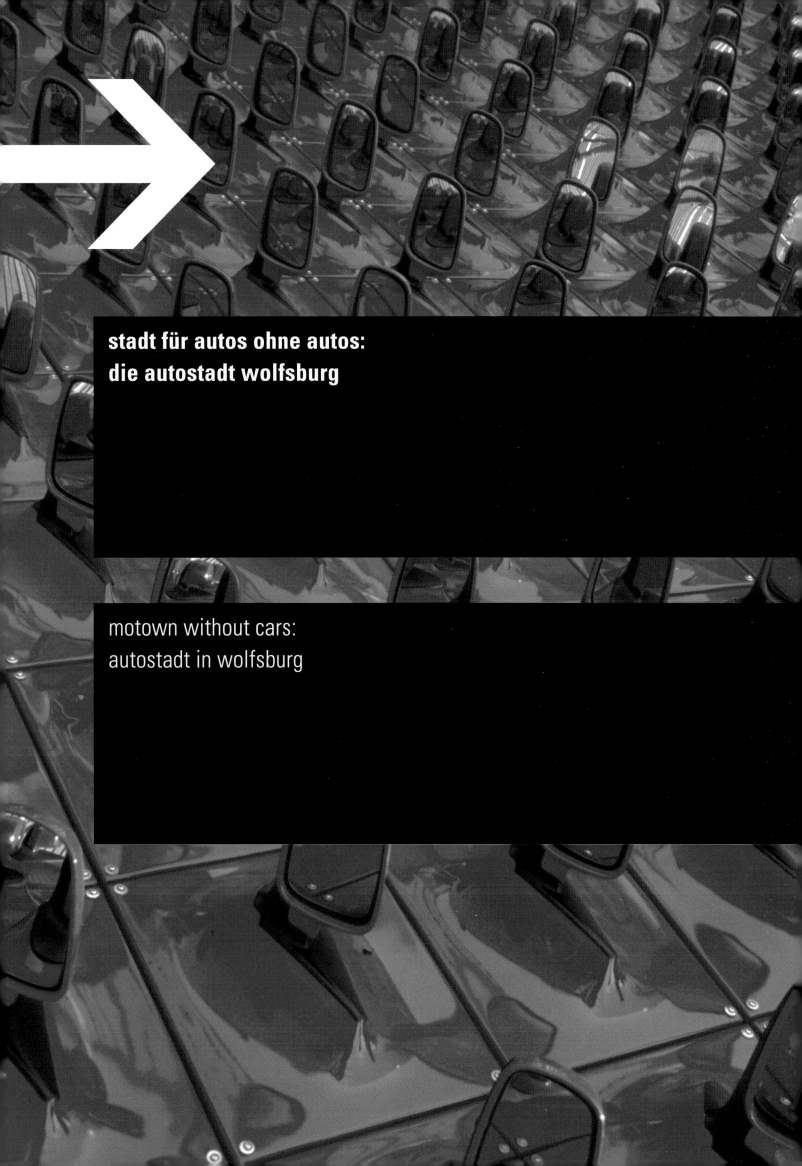

stadt für autos ohne autos:
die autostadt wolfsburg

motown without cars:
autostadt in wolfsburg

autostadt als stadtlandschaft

Die Autostadt Wolfsburg wurde zeitgleich mit der EXPO 2000 am 1. Juni 2000 eröffnet. Volkswagen betrachtet diese Stadt als „Center of Excellence" und will mit ihr in einer einzigartigen, komplexen Konstellation Service-Leistung und Kompetenz des Unternehmens zelebrieren. Von der Idee bis zur Eröffnung waren nur knapp vier Jahre vergangen. Vorher schon hatte sich für Volkswagen die Notwendigkeit verstärkt, die Auslieferung der Automobile an die Direktabholer in Verbindung mit einer kleinen Zeremonie oder Werksbesichtigung neu zu gestalten.

Zur Verfügung stand ein aufgegebenes Gewerbe- und Lagerareal am Mittellandkanal, das allerdings direkt gegenüber dem neuen Wolfsburger ICE-Bahnhof liegt und auch einen intelligenten Lückenschluss zwischen VW-Werk und Wolfsburger City zulässt. Dazu kommt, dass sich die neue Autostadt in eine gedachte Blickachse zwischen Bahnhof und Wolfsburger Schloss eingliedert.

Was als überschaubare und geniale Idee begonnen hatte, entwickelte sich zum Prototypen eines neuen architektonischen Denkens für das Automobil des 21. Jahrhunderts. Die Übergabe des Autos an seinen Besitzer wurde zur inszenierten Kult(ur)tat, bekam einen kultigen Ort, hier gar eine eigene

Stadt mit einer Corporate Architecture im Sinne der Corporate Identity von Volkswagen. Von nun an heißt es bei allen großen Unternehmungen dort, wo es um Automobilität und Architektur geht, „Lernen von Wolfsburg".

Der Name „Autostadt" ist dabei nicht nur Programm, sondern auch noch übermäßig korrekt. Die Baulichkeiten dieses neuen „Themenparks" sind eingebettet in eine neue Kunstlandschaft aus Parks, Wasser und architektonischen Statements. Eine ausgefeilte Variation der aufgelockerten modernen Stadt aus der Zeit der klassischen Moderne ist entstanden, wo die Bauten als wunderbar gestaltete Solitäre wie in einem englischen Landschaftsgarten in der in diesem Fall norddeutschen Sonne liegen.

Nur ist hier im Gegensatz zu den historischen Stadtutopien der Moderne nicht das Auto der verbindende Verkehrsträger, sondern aufgrund der sehr kurzen Entfernungen dominiert der Fußgänger: eine Autostadt ohne Autos (die befinden sich hinter Glas, wie die Juwelen beim Juwelier). Oder wie es der Chefarchitekt Gunter Henn als Überschrift über einen seiner Aufsätze setzte: „Der öffentliche Raum im Unternehmen, das Unternehmen im öffentlichen Raum". >

autostadt as townscape

Autostadt (motown) Wolfsburg was opened on 1 July 2000, at the same time as EXPO 2000. Volkswagen regards the town as a "Centre of Excellence", and this unique, complex constellation as a celebration of the company's services, achievements and competence. It only took just under four years to move from initial idea to opening ceremony, and even before that time Volkswagen had felt a pressing need to redesign the way cars were handed over to customers buying them direct ex works, combining it with a small ceremony or a factory visit.

An industrial estate directly on the Mittelland Canal served as a site which, although abandoned, was directly opposite Wolfsburg's new high-speed train station and allowed the gap between VW factory and Wolfsburg town centre to be bridged intelligently. At the same time, the new Autostadt is located right on the imaginary axis linking Wolfsburg's station with its palace.

What began as a simple stroke of genius has developed into the prototype of a new architectural concept for the car of the 21st century. The handover of a car to its owner has been transformed into a stage-managed cult(ural) act, into a cult venue and, indeed, in this case into a town of its own with a corporate

architecture compatible with Volkswagen's corporate identity. From now on, whenever the issue is one of automobility and architecture, any large company has to "learn from Wolfsburg".

As a name, "Autostadt" is not only a manifesto, but also more than appropriate. The edifices in this new "theme park" are embedded in a new artscape comprising parks, water and architectural statements. What has emerged here is a perfected variation on the theme of the heterogenous modern city from the age of classical modernism, its buildings basking in the north German sun like wonderfully designed gems in an English landscape garden.

Here, however, unlike the historical citytopias of modernism, the car is not the means of transport holding it together. Rather, the very short distances mean that it is pedestrians who are the dominant feature: a car city without cars (they are only displayed behind glass, like gems at the jeweller's). Or, to use the title given by chief architect Gunter Henn to one of his essays: "public space in the company, the company in public space". >

Schon an den Rändern zum öffentlichen Raum teilt sich die grundsätzliche Architektursprache der Autostadt mit: offen, transparent, konstruktiv und modern.
Even where it borders on public space, the fundamental architectural language of Autostadt proclaims itself: open, transparent, positive and modern.

Ein Park für Autos zum Nutzen und Vergnügen der Besucher, ganz im Sinne des klassizistischen Gartenarchitekten und -künstlers Peter Lenné: die Autostadt Wolfsburg.

A park for cars, to be used and enjoyed by visitors, just as the classicist landscape architect and artist Peter Lenné would have intended it: Autostadt Wolfsburg.

→ stadt für autos ohne autos   motown without cars

Nichts ist schwieriger als die Konzeption einer neuen Stadt – auf der grünen Wiese. Oder gar auf der Schlackenhalde. Hier bot sich die Chance, an die gigantische Linearstruktur (1) des alten Werks (von 1938) anzudocken. Die Architekten tun dies, sie formulieren ihrerseits mit den Zugangs- und Verwaltungbauten eine Zeile am Kanal (2), setzen Schwerpunkte mit sehr geometrischen Baukörpern (Halbkreis für das Ritz-Carlton-Hotel) (3), Autoregaltürmen wie einst in Chicago (4) und schließlich mit einem Doppelbaukörper am Kanal, dessen eindeutige Anleihen bei Schiffskörpern sich aufdrängen.

Diese artifiziell aufgebaute „Begrüßungsgeste" mit einer die einzelnen Baukörper verbindenden Piazza (5) und eine anschließende geschickte Durchwegung des Parks bilden den Rahmen, oder, wenn man will, eine Art

Handmulde für die fröhliche Aufstellung der einzelnen (Ausstellungs-) Pavillons im Innern der Autostadt. Alle wichtigen Volkswagenmarken haben hier ihr eigenes Zuhause: Volkswagen (6), Audi (7), VW-Nutzfahrzeuge (8), Seat (9), Bentley (10), Lamborghini (11), Skoda (12).

In diesem Bereich wird die Autostadt zur Gartenstadt, hier sind es Hügel, Grün und Wasser, die die Architektur inszenieren, die Bühne zur perfekten Performance bilden; ein gekonntes, aber trainiertes Spiel, das den Augen zugute kommt. Nichts wirkt hier streng, nichts aufgezwungen, alles ist so, wie der große deutsche Landschaftsarchitekt Peter Lenné es immer anstrebte und auch erreichte: neben dem Nutzen wird auch an das Vergnügen für die Parkbesucher gedacht (Gestaltung: WES & Partner, Hinnerk Wehberg).

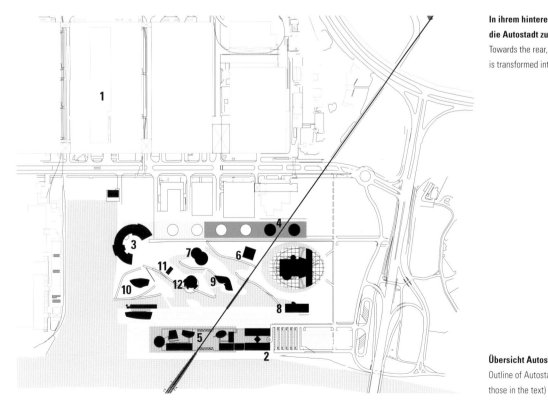

In ihrem hinteren Teil, wo die Markenpavillons liegen, wird die Autostadt zur schwingenden Grünlandschaft.
Towards the rear, where the marques have their pavilions, Autostadt is transformed into a rolling park landscape.

Übersicht Autostadt (Ziffern vgl. auch Text)
Outline of Autostadt (numbers correspond to those in the text)

Autos sind nicht zugelassen, dafür eher Boote: kleiner See mit Seat-Pavillon.

Cars are not allowed in, but boats have a better chance: small lake and Seat pavilion.

Alt und neu: Das VW-Werk aus Vorkriegszeiten wird mit einer modernen Pavillonarchitektur (hier Bentley-Pavillon) konfrontiert.

Old and new: The pre-war VW plant contrasts with modern pavilion architecture (the picture shows the Bentley pavilion).

There is nothing more difficult than to design a new town – on a greenfield site. Or indeed on a slag heap. In this case, there was an opportunity to dock onto the gigantic linear structure (1) of the old plant (built in 1938). This is what the architects do, themselves drawing a line along the canal with the entrance and administrative buildings (2), setting accents with extremely geometrical structures (semi-circle for the Ritz Carlton Hotel) (3), car rack towers reminiscent of Chicago (4), and finally with a double structure along the canal, with its striking, undisguised allusions to ships' hulls.

This artificially constructed "welcome gesture", with a piazza linking the individual structures (5) and a subsequent, ingenious system of paths crossing the park, form the setting – indeed, one could almost say a kind of hollow –

for the gay arrangement of individual (exhibition) pavilions at the heart of Autostadt. Each major Volkswagen marque has a home of its own here: Volkswagen (6), Audi (7), VW commercial vehicles (8), Seat (9), Bentley (10), Lamborghini (11), Skoda (12).

In this area, Autostadt becomes a garden city, with hills, green spaces and water setting the scene for architecture, creating the stage for a perfect performance; a masterly though not unrehearsed play that is pleasing to the eye. There is nothing here that seems severe or affected, everything is just as Peter Lenné, the great German landscape architect, always aimed for and indeed achieved: he wanted his park visitors to have pleasure as well as utility (Design: WES & Partner, Hinnerk Wehberg).

konzernforum und piazza: die offene tür

Es ist mehr als eine Geste: In der Autostadt wird der Bahnreisende hofiert, wenn er aus dem ICE steigt. Mit dem KonzernForum am anderen Ufer des Mittellandkanals als Orientierungsmarke fest im Blick wird er magisch von diesem gläsernen Zeichen angezogen. Das Bauwerk birgt eine beeindruckende Lobby, die so manche Opern- oder Theaterhalle in den Schatten stellt.

Von hier aus – alles ist überdacht – werden die Besucherströme zum Empfang, in die Gastronomie und in die Hallen der KonzernWelt geleitet. Die Piazza ist das Bindeglied: Der öffentliche Raum der Stadt geht nahtlos

in die Autostadt über – wegen des unberechenbaren Wetters liegt alles unter weitem Dach. Durchschreitet der Besucher die Halle Richtung Park, erreichen ihn die einzelnen Ausstellungsteile, ausgewogen und doch überraschend arrangiert, wie Kulissen auf einer spannenden Bühne.

Konstruktion/Materialien ———— 45 m Rohrprofilfachwerkbinder und Stahl-Verbundstützen als Dachkonstruktion
Architekten ————————— Henn Architekten, München
Kunst ————————————— Rot Blau Gelb von Gerhard Merz, Globen von Ingo Günther

Haushohe Glaspaneele öffnen sich einladend: Zutritt ohne Schwellenangst.
Glass panels as high as the building open invitingly: this threshold can easily be crossed.

Das KonzernForum gesehen aus der Autostadt.
KonzernForum seen from Autostadt.

In der mächtigen überdachten Lobby (Piazza) werden die Besucherströme empfangen und verteilt.
Under the cover of a roof in the massive entrance hall (piazza), the streams of visitors are welcomed and directed to their destinations.

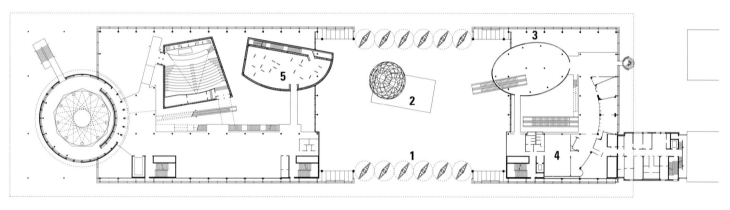

1 Haupteingang	2 Piazza	3 Kundenempfang	4 Restaurants	5 Ausstellung, Kino etc.
1 Main entrance	2 Piazza	3 Customer reception	4 Restaurants	5 Exhibition, cinemas etc.

konzernforum and piazza: the open door

In what is more than a gesture, the rail traveller is wooed in Autostadt as soon as he descends from the bullet train. With his eyes firmly fixed on the landmark of the KonzernForum on the opposite bank of the Mittelland Canal, he is magically attracted by this glass showcase. On the inside, the building has an impressive entrance hall that puts many a theatre or opera house foyer to shame.

From here, the streams of visitors are guided along covered walkways to the reception area, to eating areas and to the halls of this ingenious corporate world. It is the connecting element: the transition from the public space of the town to Autostadt is a smooth one – and because weather is so unpredictable, everything is contained under a sheltering roof. Crossing the foyer and heading towards the park, the visitor comes to the individual parts of the exhibition, arranged harmoniously yet surprisingly, like props on a fascinating stage.

Construction/materials	45-metre tubular section trusses und composite steel supports as roof structure
Architects	Henn Architekten, Munich
Artworks	Rot Blau Gelb (red blue yellow) by Gerhard Merz, Globes by Ingo Günther

zeithaus: das duale automuseum

Es ist wahrscheinlich das erste Automuseum, das mit seinen beiden Haushälften darauf aufmerksam machen will, dass unsere Welt zweigeteilt ist: in Virtualität und Realität, in Anfassbares und Vorstellbares. Anfassbar sind in diesem Automuseum die Autos, bewährt in einem gläsernen Regal konzentriert. Auf der anderen Seite des „gap", einer Schlucht für Rolltreppen und Treppen, stehen mitunter zwar auch Autos, aber hier herrscht eine Inszenierung der kulturellen und sozialen Flora rund um das Automobil vor. „Memory Lane" nennen das die Ausstellungsmacher, und sie sprechen auch von den zwei Hälften eines Gehirns. Vielleicht sind es auch Yin und Yang oder Max und Moritz, in jedem Fall ist es eine perfekt aufgefädelte Show ohne Leerlauf.

Konstruktion/Materialien ———— Das nördliche Glasregal ist ein Stahlskelettbau; die
Treppenhalle ist teilweise in die Dachkonstruktion gehängt
Architekten ——————— Henn Architekten, München

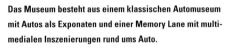

Das Museum besteht aus einem klassischen Automuseum mit Autos als Exponaten und einer Memory Lane mit multimedialen Inszenierungen rund ums Auto.
The museum consists of a classical car museum and a Memory Lane with multimedia productions centred on the car theme.

zeithaus: the dual car museum

With its two building halves, this is probably the first car museum that tries to draw our attention to the dichotomy of our world: the virtual and the real, the tangible and the imaginary. The cars are what are tangible in this car museum, concentrated conventionally on a glass rack. On the other side of the "gap" – a chasm for stairways and escalators – there are cars too, but here the emphasis is on a dramatic production of the cultural and social flora surrounding the car. The exhibition makers call it "Memory Lane", and also speak of the two halves of the brain. They may also be Yin and Yang or Tom and Jerry, but at all events this is a perfectly strung together show which never flags.

Construction/materials ———— The north-facing glass rack is a steel skeleton construction;
the stairwell is partly suspended from the roof structure
Architects ——————— Henn Architekten, Munich

Oben und rechts: Der südliche, geschlossene Teil des ZeitHauses nimmt die Memory Lane auf.
Top and right: The south-facing, closed part of the ZeitHaus houses the Memory Lane.

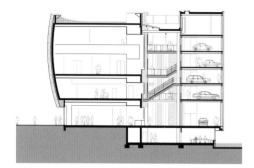

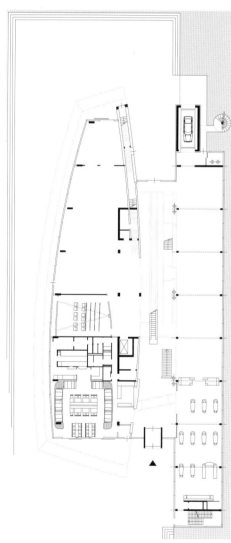

Schnitt und Grundriss
Section and plan

Eine VW-Ikone in der Memory Lane: der Käfer.
A VW icon in Memory Lane: the Beetle.

**Memory Lane und Automuseum werden durch eine
Art Schlucht getrennt. Dort liegen Rolltreppen und Treppen
(links und ganz links).**
Memory Lane and car museum are separated by a gap with
escalators and steps (left and far left).

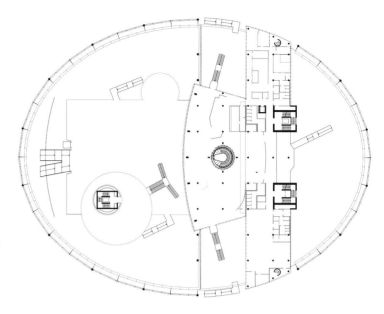

im herzen der stadt: kundencenter und autotürme

In der lichten Weite eines elliptischen Baukörpers befindet sich das eigentliche Herz der Autostadt. In einer vollkommenen Inszenierung erreichen die Autabholer über eine breite Showtreppe eine Lobby mit der Großzügigkeit eines Airport-Terminals. Auf großen Displays erfahren die Käufer die Daten ihres Rendezvous mit ihrem neuen „Familienmitglied", ein Geschoss tiefer warten die Autos auf die Übergabe. Elemente einer offenen technoiden Architektur wie im Eingangsgebäude bilden die beeindruckende Kulisse: Abbild der Perfektion eines Unternehmens, das 100 Millionen Fahrzeuge und mehr gebaut hat.

Direkt zugeordnet sind die beiden AutoTürme, die ca. 400 Fahrzeuge aufnehmen können. Diese werden über Fahrstühle zur Übergabe befördert. Etwa alle 40 Sekunden wird unterirdisch aus dem benachbarten Werk ein Auto in die Türme geliefert.

Konstruktion/Materialien ——— **Türme: feuerverzinkte Stahlkonstruktion,**
KundenCenter: An einem zentralen Pylon aufgehängte Konstruktion, die weitgehend deswegen stützenfrei ist.
Architekten —————— **Henn Architekten, München**

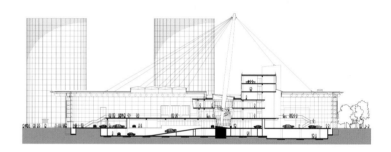

in the heart of the city: kundencenter and autotürme

The real heart of Autostadt is situated in the wide open space of an elliptic structure. In a perfect production, customers who have come to collect their cars ascend a broad show staircase and arrive in a foyer as colossal as an airport terminal. On large-format displays, customers can read the data relating to their rendezvous with their new "family member", while one floor below cars wait to be handed over. This impressive setting comprises the same elements of open, technoid architecture as the entrance building, and reflects the perfection of a company that has built 100 million vehicles or more.

The building is directly linked to the two AutoTürme (car towers), which can hold roughly 400 vehicles. Via lifts, the cars are taken to the handover point. Roughly every 40 seconds, a car is taken underground from the neighbouring plant to the towers.

Construction/materials ————— Towers: galvanized steel structure.
KundenCenter: construction suspended from a central tower, making it largely support-free
Architects ————————— Henn Architekten, Munich

Oben: Grundriss KundenCenter
Mitte und unten: Schnitt und Ansicht KundenCenter und AutoTürme
Top: plan of KundenCenter
Centre and bottom: Section and general view of KundenCenter and AutoTürme

Glas und Metall sind die Leistungsträger für die Architektur des VW-Pavillons, der AutoTürme und des KundenCenter (von links).

Glass and metal are key players in the architecture of (from left to right) the VW pavilion, the AutoTürme and the KundenCenter.

Sie warten hier auf ihre neuen Besitzer und Abholer: die fertigen Autos aus Wolfsburger Produktion. Die Banner im KundenCenter stammen von Matt Mullican.

Waiting for their new owners to come and collect them: finished cars from the Wolfsburg factory. The banners in the KundenCenter are by Matt Mullican.

**Die AutoTürme nehmen
bis zu 400 PKW auf, alle
40 Sekunden kann ein
Neuwagen aus dem Werk
zugeliefert werden.**
The AutoTürme hold up to
400 cars, a new car can
be added from the plant every
40 seconds.

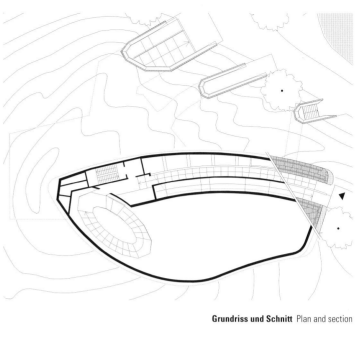

bentley-pavillon: der berg ruft

Dieser Pavillon liegt im Südwesten des Geländes zwischen ZeitHaus und dem Hotel Ritz-Carlton. Wenn Bentley im VW-Konzern eine der Topmarken neben Bugatti und Lamborghini ist, also die Krone bildet, so gilt das auch für die Phantasie bei der Architektur des Pavillons. Er hat etwas von einer Schatzkammer oder einem Schrein. Denn der Großteil des Volumens steckt in der Erde, der Zugang ist kurvig und wirkt wie die Miniatur einer Auto-Rennstrecke. Der Besucher wird von der sich zum Himmel öffnenden Vitrine praktisch eingesogen. Er folgt dann einem spindelförmigen Besuchs-Parcours um die gigantische Kopie einer Bentley-Kurbelwelle herum. Die auffallende symbolische Leistung des Entwurfs entführt den Besucher aus dem Alltag in die Tiefe des Fertigungsgeheimnisses einer absoluten Kultmarke.

Grundriss und Schnitt Plan and section

Besucherkapazität		900 pro Stunde
Ausstellungsfläche		650 qm
Materialien		Glas, Stahl, grüner Granit
Architekten		Henn Architekten, München und KSS, London

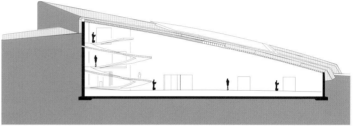

**Der Parcours führt spindelförmig um die 8 m lange
Nachbildung einer Bentley-Kurbelwelle herum.**
The visitors' trail winds its way around the eight metre-
high copy of a Bentley crankshaft.

bentley pavilion:
the mountain beckons you to discover its innermost secrets

This pavilion is situated in the south-west of the site, between ZeitHaus and
Hotel Ritz-Carlton. If Bentley is one of the leading marques in the VW Group,
alongside Bugatti and Lamborghini, the crown as it were, this is also the
case as regards the imaginative architecture of the pavilion. It has something
of the appearance of a treasury or a shrine. Most of the pavilion's mass is underground, and the approach to it is curved like a miniature racetrack. The visitor
is sucked in by this show case, which opens out to the sky. He then follows a
guided trail which winds its way around the gigantically dimensioned copy of a
Bentley crankshaft. What is so striking and symbolic about this design is the way
it whisks the visitor away from the everyday world to the profound secrets
of the manufacture of an absolute cult marque.

Maximum number of visitors	900 per hour
Exhibition space	650 m^2
Materials	Glass, steel, green granite
Architects	Henn Architekten, Munich and KSS, London

**Die Schornsteine des VW-Werks bewachen eine Schatzkammer
zur Präsentation britisch-deutscher Autobaukultur.**
The chimneystacks of the VW plant watch over a treasure chamber
designed to present British/German car manufacturing culture.

vw und audi: markenzeichen werden merkzeichen

Zeitlosigkeit und Perfektion – bei VW und Audi werden die entsprechenden Markensignets zum Ideen- und Taktgeber der Architektursprache. Bei VW umschließt ein gläserner Würfel eine Kugel, die ein 360°- Kino beinhaltet. In der Ansicht erinnert die Komposition an das VW-Zeichen. Für die Macher selbst ist die Kugel im Würfel „Sinnbild für absolute Perfektion". Präziser ist Corporate Identity selten architektonisch umgesetzt worden. >

VW-Pavillon

Besucher	**700 pro Stunde**
Ausstellungsfläche	**748 qm**
Materialien	**Würfel: reine Stahlkonstruktion, Doppelfassade**
	mit 2000 einzeln steuerbaren Lamellen
	Kugel: doppelwandige Stahl-/Metallkonstruktion
Architekten	**Henn Architekten, München und Grüntuch und Ernst, Berlin**

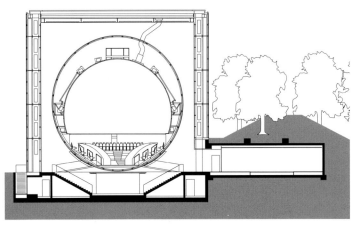

vw and audi: from brand name to distinguished feature

Timelessness and perfection – at VW and Audi the respective logos inspire and drive the architectural language of their two pavilions. In the case of VW, a glass cube envelops a globe containing a 360° degree cinema. The impression that results from this composition is that of the VW logo. For the makers themselves, a globe in a cube is a "symbol of absolute perfection". Rarely has the embodiment of corporate identity been architecturally more precise. >

VW pavilion

Visitors	700 per hour
Exhibition space	748 m^2
Materials	Cube: pure steel construction with double façade,
	with 2000 individually controllable louvres
	Globe: double-walled steel-metal structure
Architects	Henn Architekten, Munich and Grüntuch und Ernst, Berlin

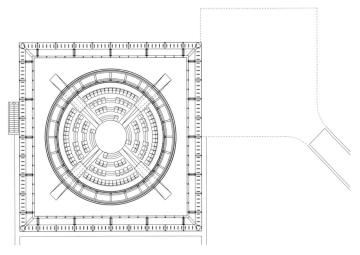

Schnitt und Grundriss Section and plan

Kugel im Würfel oder Kreis im Quadrat als „Sinnbild für absolute Perfektion".
A globe in a cube or a circle in a square: the "symbol of absolute perfection".

Beim Audi-Pavillon sind im Grundriss vier ineinander greifende Kreise zu erkennen; natürlich ist das eine Anspielung auf die vier Ringe des alten Auto-Union-Konzerns mit seiner Marke Audi, die jetzt zum VW-Konzern gehört. Der Besucher wird dieses zweidimensionale Spiel kaum realisieren, für ihn ist die Liaison zwischen Design und Technik offensichtlicher. Hier bleibt die Präsentation der Markenphilosophie nicht beim Medium Film stehen, sondern wird zu einem Kleinstmuseum entlang einer so genannten Kreationsspirale (vgl. auch Audi Forum Ingolstadt, S. 168 ff).

Wenn beim Audi-Pavillon das schwebend elegante Fassadendesign festgelegt ist, wird es bei VW zu einer sich immer wieder ändernden Installation: Die beweglichen Lamellen in der Fassade können höchst unterschiedlich konfiguriert werden. Das Innere der Kugel soll nicht für alle Zeiten als Kino genutzt werden. Veränderungen sind geplant, der Raum ist flexibel genug dafür.

The layout of the Audi pavilion reveals four intermeshing circles, which are of course an allusion to the four rings of the former Auto-Union group and its Audi marque, which is now part of the VW group. Visitors are scarcely aware of this two-dimensional device, but what they do see is the liaison between design and technology. Here, film is not the only medium used to present the brand philosophy; rather, the pavilion is a miniature museum following a "creative spiral", (see also the Audi Forum in Ingolstadt, p. 168 ff).

While the floating, elegant design of the Audi pavilion façade is fixed, in the case of VW it is an ever changing installation: the movable louvres in the façade can be configured in any variety of ways. It is not intended to use the interior of the globe as a cinema for ever. Changes are planned, and the space is flexible enough to accommodate them.

Audi-Pavillon

Besucher	500 pro Stunde
Ausstellungsfläche	1.400 qm
Materialien	koventionelle Konstruktion
Architekten	Henn Architekten, München

Audi pavilion

Visitors	500 per hour
Exhibition space	1,400 m^2
Materials	Conventional structure
Architects	Henn Architekten, Munich

Links: Großzügig in den Park plantiert: die Ringe des Audi-Pavillons.
Rechts: Die Ausstellung ist spiralförmig angeordnet.
Left: in a generous park setting: the rings of the Audi pavilion.
Right: the exhibition is arranged in a spiral.

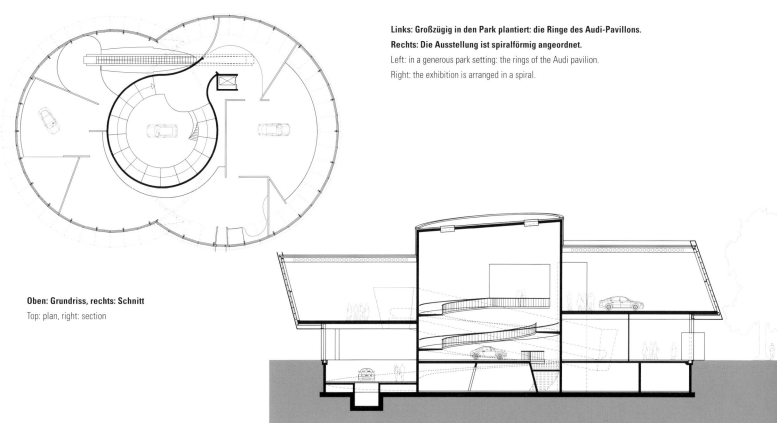

Oben: Grundriss, rechts: Schnitt
Top: plan, right: section

raubtierkäfig und spanische lust: seat und lamborghini

Zwei weitere Marken zeigen die ganze Bandbreite des Konzerns und der möglichen Architektur auf: Lamborghini wird zur rotierenden Installation ungebändigter Kraft, präsentiert sich als „geschlossener Käfig des ungezähmten Raubtiers", dem Sportwagen.

Seat setzt auf spanische Kreativität an der Schnittstelle zwischen Produktdesign und Architektur. Dieser Pavillon soll wie eine „sinnlich geschwungene, strahlend weiße Skulptur wirken" und an die Ramblas in Barcelona erinnern.

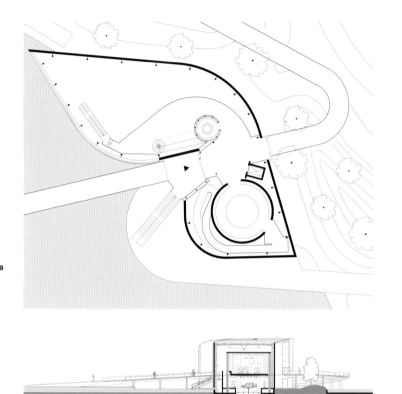

Seat-Pavillon

Besucher	600 pro Stunde
Ausstellungsfläche	1.100 qm
Architekten	Henn Architekten, München und Alfredo Arribas, Barcelona

Lamborghini-Pavillon

Besucher	600 pro Stunde
Ausstellungsfläche	311 qm
Architekten	Henn Architekten, München und Bellprat, Winterthur

Seat bei Nacht. Rechts: Grundriss und Schnitt Seat-Pavillon
Seat at night. Right: Plan and section of the Seat pavilion

Lamborghinis Raubtierkäfig
Lamborghini's predators cage

Seat-Eingang mit Autospiegelkomposition
Seat entrance with car mirror composition

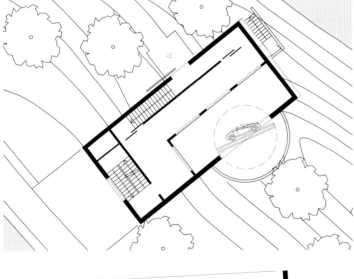

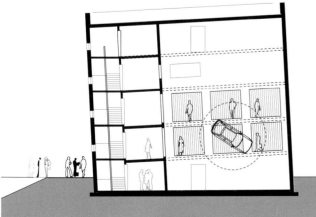

predators cage and spanish zest: seat and lamborghini

Two other marques neatly demonstrate the breadth of the group and of possible architectural solutions. On the one hand, there is Lamborghini, whose rotating installation of untarned power presents a "wild predator kept behind bars", the sports car.

Seat, by contrast, goes for Spanish creativity at the interface between product design and architecture. This pavilion is intended to have the effect of a "sensually curved, gleaming white sculpture", reminiscent of Barcelona's ramblas.

Seat pavilion

Visitors	600 per hour
Exhibition space	1,100 m^2
Architects	Henn Architekten, Munich and Alfredo Arribas, Barcelona

Lamborghini pavilion

Visitors	600 per hour
Exhibition space	311 m^2
Architects	Henn Architekten, Munich and Bellprat, Winterthur

Grundriss und Schnitt Lamborghini-Pavillon
Plan and section of the Lamborghini pavilion

interieurs und restaurants: design der extraklasse

Zum Gestaltungskonzept der Autostadt gehörte auch ein generelles Arbeitsteilungsgebot. Gunter Henn als Master Planer und Entwurfsarchitekt ging dort, wo es sinnvoll war, jeweils Allianzen mit Spezialisten und Architekten aus dem Lande der jeweiligen Marke ein, wie Confino & Co, KSS Architects, Bellprat Associates, Alfredo Arribas Arquitectos Asociados, Studio Borek Sipek, Andrée Putman, Toni Chi & Associates, Jordan Mozer & Associates oder Virgile & Stone Associates. Ein Beispiel des Architekten und Designers Jordan Mozer („Räume müssen lebendig und menschlich sein") beweist, wie eine solche Strategie die Bandbreite der Gestaltung wohltuend vergrößert. Mozer hat mit dem „Cylinder" im ZeitHaus ein American Diner neu erfunden, es ist hier eine „Interpretation des Mythos Auto. Die Architektur verkörpert Beschleunigung und Abenteuer" (Jordan Mozer). Hier werden Mobiliar, Wand und Decke zum Abbild eines Auto-Interiors.

Rechts: Rutsche im „Auspuff", Kinderhaus im „Motorblock": Die Welt für die ganz kleinen Autofahrer in der Autostadt. Oben: Das „Cylinder" versetzt den Besucher in die Welt eines amerikanischen Autostrips.

Right: a slide in the exhaust, a Wendy house in the engine block: a world for very young car drivers at Autostadt. Top: the Cylinder Restaurant transports the visitor to a world of American car strips.

interiors and restaurants: outstanding design

The design concept for Autostadt also included a general principle of division of labour. As master planner and the planner responsible for the overall design of the buildings, Gunter Henn entered into alliances with specialists wherever this made sense, or with architects from the brands' respective countries of origin, such as Confino & Co, KSS Architects, Bellprat Associates, Alfredo Arribas Arquitectos Asociados, Studio Borek Sipek, Andrée Putman, Toni Chi & Associates, Jordan Mozer & Associates or Virgile & Stone Associates.

An example provided by the architect and designer Jordan Mozer ("rooms must be human, and full of life") shows how good such a strategy is for enhancing the band-width of design. With his "Cylinder Restaurant" in the ZeitHaus, Mozer has reinvented an American diner, making it into an "intrepretation of the car myth. Its architecture expresses acceleration and adventure in tangible form" (Jordan Mozer): Here, furniture, walls and ceiling are like the interior of a car.

Gibt es eine bestimmte, also eine adäquate Architektursprache, wenn man für das Auto Häuser baut?

Gebäude, die Bewegung und Abläufe sichtbar machen, entsprechen wahrscheinlich am ehesten der Dynamik eines Autos. Deswegen bauen wir keine Räume, die sich verschließen, sondern sich öffnen und ineinander fließen.

Das hat zu einer lichten und transparenten Glas-Stahl-Architektur geführt, die schon wegen ihrer Oberflächen eine Affinität zur modernen Industriearchitektur offenbart. Kann man dann optisch noch zwischen Architekturkonstruktion und Werkzeug unterscheiden?

Es geht nicht um den architektonischen Raum in seiner ästhetischen Bedeutung, sondern darum, mit der Architektur die Abläufe der Produktion deutlich zu machen; und es gibt keine richtige Schnittstelle zwischen Architektur und Werkzeug. So werden dann Aufzüge, Gehänge und Skips zum integralen Bestandteil des gesamten Raumeindrucks.

Es sieht so aus, als ob mit der Dresdner Manufaktur ein Quantensprung bei der Planung von Autofabriken erreicht wäre. Nicht verwunderlich, weil Sie auf Erfahrungen mit der erfolgreichen Autostadt und Skoda zurückgreifen können. Haben Sie den Typus einer neuen urbanen und integrierten Autofabrik gebaut?

Wir sind dabei, eine „Charta von Dresden" zu schreiben. Das meine ich in Anspielung auf die berühmte „Charta von Athen", in deren Folge die Fabriken

aus der Stadt und den Wohngebieten vertrieben wurden. Nur ist die Charta von Dresden ein Umkehrschluss: Die Industrie kehrt zurück, weil sie kleinteilig, mit dem Charakter von Handwerksbetrieben, wieder stadtfähig geworden ist. Handwerksbetriebe für Kutschen oder Möbel lagen immer mitten in den Städten. Wir schauen doch gern einem Tischler bei der Arbeit zu, wollen doch wissen, wie etwas hergestellt wird, besonders, wenn es hochwertig ist. Und die Menschen hier, Besucher wie Angestellte, merken, dass mit der Fertigung des Phaeton eine hochwertige Leistung vollbracht wird.

Sie haben einen neuen Kultort für das Auto gebaut?

Die Gläserne Manufaktur ist tatsächlich ein Ereignisraum, der das kulturelle Reservoir an historischen Sehenswürdigkeiten Dresdens nutzt und komplettiert. Die Stadt wird um einen neuen identitätsstiftenden Ort bereichert.

Über Jahrhunderte hinweg bis heute zählen Sakralbauten zu den Lieblingsaufgaben von Architekten. Sie haben in einem Vortrag davon gesprochen, dass das Bauen für und rund um das Auto etwas Sakrales besitzt …

Es hängt damit zusammen. Architektur, d.h. Gebäude werden durch zwei Dinge bestimmt: den Raum und den Baukörper. Gebäude sind Transformationen des Raumes durch Baukörper. Sie brauchen das Materielle, den Baukörper, den sie dann durch Materialien und Farben genauer bestimmen. Aber es entsteht ein Raum. Und es ist immer das Interesse eines Architekten, einen Raum nicht nur als Begrenzung zu haben, sondern mit dem Raum

→ **interview** interview

Gunter Henn, der Prinzipal des Münchner Büros Henn Architekten, ist in den letzten Jahren zum wichtigsten Architekten für den Volkswagen-Konzern avanciert und hat sich auch in dieser Rolle stark mit dem Thema Auto und Architektur auseinander gesetzt.
Gunter Henn, the head of the Munich-based Henn Architekten bureau, has progressed over the last few years to become Volkswagen AG's leading architect. In this role, he has taken a long, critical look at the topic of cars and architecture.

Is there a specific, or shall we say appropriate, architectural language when designing buildings for the motor car?

Buildings that make movement and processes obvious probably come closest to the dynamism of the car. This is why we do not build any rooms that shut themselves off, but rather rooms that open out and flow into each other.

Frequently, the result of this has been a light, transparent architecture of glass and steel, whose surfaces alone reveal an affinity with modern industrial architecture. Once this happens, is it any longer possible to distinguish between architectural design and utility?

The issue here is not one of architectural space in its aesthetic sense, but of using architecture to make the processes of production apparent; and there is no longer any real interface between architecture and utility. In this way, lifts, suspension gear and skips become an integral part of the overall architectural impression.

It would seem that the Dresden factory has achieved a quantum leap in the planning of car factories. This is not surprising, given that you were able to draw on your experience with the successful Autostadt and Skoda. Have you built a building type that is an urban, integrated car factory?

We are in the process of drafting a "Charter of Dresden". I say this as an allusion to the famous "Charter of Athens", which resulted in factories being driven

out of cities and their residential areas. The Charter of Dresden, however, turns this argument on its head: industry is returning, because it has become fragmented like artisanal workshops, and has thus become compatible with the urban environment again. Artisanal workshops for carriages or furniture were always to be found at cities' hearts. We enjoy watching a cabinet-maker at work, we want to know how something is made, especially if it is a precious object. And the people here, both visitors and workers, recognize that the work of manufacturing a Phaeton is an outstanding achievement.

Are you saying that you have built a new cult venue for the car?

There is no doubt that the glass factory is a venue for events which exploits and completes Dresden's cultural "reservoir" of historic sights. The city has gained a new attraction, one that contributes to its identity.

For centuries now, religious buildings have been among architects' favourite tasks. In a lecture, you once said that building for and in relation to the car has a kind of religious quality …

There is a connection. Architecture, i.e. buildings, is the product of two things: space and the building structure. Buildings are transformations of space by building structures. They need a tangible element, the building structure, which they then define more precisely by means of materials and colours. But it is a space that comes into being. And an architect is always concerned to have space not as a restricting factor, but to use this space to open up other spaces: spiritual

Räume zu erschließen: spirituelle Räume, immaterielle Räume, trotz aller Materialität. Ein Kirchenraum ist ein Prototyp dafür. Aber eine solche Aura kann man auch heute beim Bauen schaffen.

Ob man nun von Sakralität oder Aura spricht, in jedem Fall sind die funktionalen Zwänge einer Autofabrik weit größer als im Kirchenbau ...

Richtig. Aber auch bei solchen sehr funktional geprägten Bauaufgaben wie einer Automobilfabrik geht es mir um „Aura-Qualität", die „awareness of quality". Es geht nicht nur um eine Autoproduktion, sondern beispielsweise um die Aura eines Qualitätsbewusstseins. Und das wird bei Skoda spürbar, und vor allem bei der Produktion des Phaeton in Dresden.

Was ist „awareness" beispielsweise bezogen auf die Autostadt?

Es ist mehreres. Aber man kann sagen, vor allem die „awareness of communication". Auf den Dialog kommt es an. Es gibt komplexe Zusammenhänge, die sind mit einem klassischen Marketing oder einem Werbespot nicht zu vermitteln, also Dinge wie Gesundheit, Bildung, Religion und Mobilität. Das sind keine Produkte, das sind Themen, die berühren mich in meinem Lebensstil, in meiner Lebenshaltung. Und solche komplexen Themen kann man nicht einfach mit einem Werbespot im Fernsehen behandeln, sondern hier muss man Dialogplattformen schaffen. Da muss man narrativ arbeiten, muss man Geschichten erzählen. In unserem Falle heißt das Ergebnis Autostadt, und die ist eine urbane Kommunikationsplattform.

Welche Rolle spielt die gewaltige Industriekulisse des VW-Werks für die Autostadt?

Eine sehr wichtige im Sinne von Authentizität. Die Autostadt liegt nicht auf der grünen Wiese, dann wäre sie bloß ein Themenpark. Davon gibt es viele. Die Autostadt wird geführt von der Organisation, von der Logistik der Autofabrik und deshalb ist es ganz wichtig, das Kraftwerk sehen zu können. Man spürt, dass dahinter das größte Automobilwerk der Welt liegt. Sehr, sehr viele Autostadt-Besucher machen eine Tour durch die Fabrik. Es ist sicher die größte Attraktion, durch ein Presswerk und eine echte Autofabrikation zu fahren. Denn dann hat man die Gewissheit, dass das Auto, das man abholt, auch dort produziert wurde. Wir wollen wissen, wo die Tomate herkommt, aus Holland oder von den Kanarischen Inseln. Diese Authentizität ist ein ganz wichtiges Element, wie in Dresden.

Und nun planen Sie eine AutoUniversität in Wolfsburg. Was wird das sein können?

Es geht bei der AutoUniversität nicht um das akademische Wissen, das anerkannt ist und einfach zu vermitteln ist. Es soll ein Ort geschaffen werden, wo Wissen gelehrt wird, aber dieses Wissen in diesem Moment erst entsteht. Es ist ein Ort, an dem vier Wissenscluster zusammenkommen: Mobilität, das ist die Kompetenz von Volkswagen. Dazu Gesundheit (Health-Care), IT (Informationstechnologie) und Dienstleistungen. Dienstleistung wird komprimiert durch die Autostadt dargestellt. IT ist heute überall notwendig, genauso wie Health-Care. Wichtig sind aber die Synergieeffekte.

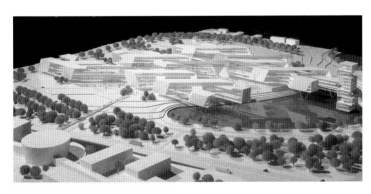

space, intangible space, despite all their materiality. The enclosed space of a church is a prototype of this. However, its aura is something that can be captured by modern architects.

We may speak of religiosity or aura, but at all events the functional contingencies of a car factory are far greater than in church-building ...

That's true. But even in building tasks that are as functionally conditioned as a car factory, my concern is one of "aura quality", and the "awareness of quality". The issue here is not one of car production, but of the aura, say, of quality consciousness. This can be felt at Skoda, and above all in the production of the Phaeton in Dresden.

And with respect to Autostadt? What does "awareness" mean there?

It is several things. But you could say that the most important one is "awareness of communication". Dialogue is the key issue here. There are thematic complexes that cannot be conveyed by classical marketing or a TV commercial; things like health, education, religion or mobility. They are not products, but topics that affect me in my lifestyle and my outlook on life. And topics as complex as these cannot simply be dealt with by a TV commercial. Rather, what we have to create here are platforms for dialogue. You have to work narratively here, you have to tell a story. In our case, the result of this is Autostadt, an urban communication platform.

What role does the overwhelming industrial backdrop of the VW plant play for Autostadt?

In terms of authenticity, a very important one. Autostadt is not simply on a greenfield site, for that would make it merely a theme park. There are plenty of those. Autostadt takes its lead from the organization and logistics of the car factory, and so it is very important that the plant can be seen. You can feel that the world's largest car factory is behind Autostadt. Very, very many visitors to Autostadt take a tour around the factory. The greatest attraction is without doubt that of riding through a pressing plant and a true car factory. For then people can be sure that the car they are collecting was in fact also produced there. We want to know where our tomatoes come from, whether they are from Holland or the Canaries. This authenticity is an extremely important element, as it is in Dresden.

And now you are planning an "AutoUniversität", a car university, in Wolfsburg. What will that be about?

The issue at AutoUniversität is not one of academic knowledge that is recognized and easy to impart. The idea is to create a place that teaches knowledge, but knowledge that has only just emerged. It is a place at which four clusters of knowledge converge. First, mobility. That's where Volkswagen comes in. Then health care, IT, and services. Services are presented in condensed form by Autostadt. IT is needed everywhere today, as is health care. What is important, though, is the synergy between these elements. This requires mobility – both physical mobility and mobility of another kind: intellectual and emotional mobility.

Dafür braucht man Mobilität – physische Mobilität und die andere, die mentale, emotionale Mobilität. In der AutoUni sollen Wissensräume entstehen, neue Werteräume, neue Sinnräume.

Es geht ja auch um die Frage, inwieweit das Thema Mobilität, Auto-Mobilität eine Perspektive hat. Was können wir mit ihr erreichen?

Das wird aufgegriffen. Das Besondere an der AutoUni sind so genannte Future-Labs. Es ist nicht nur eine Universität, sondern die Verbindung von Universität und Unternehmen, beziehungsweise der Wirtschaft. Vertreten durch Motorola, Intel, IBM, Hewlett-Packard oder die Harvard Medical School. All diese Institutionen werden dort Future-Labs mit fünf bis zehn Leuten unterhalten, Entwickler, Forscher oder Spinner, die querdenken, um festzustellen: In welche Richtung kann es laufen? Wie wird Automobilität in zehn Jahren zu verstehen sein? Wie wird physische, wie emotionale Mobilität gemacht werden?

Was bedeutet das Thema „Stadt und Auto" für Sie? Da haben wir doch beträchtlichen Forschungsbedarf?

Ich kann mir das sehr gut vorstellen und wünsche mir auch, dass in dieser Universität ein Bereich sich sehr differenziert mit der Zukunft der Stadt, Zukunft der Urbanität und der Verbindung von Urbanität und Mobilität auseinander setzt. Obwohl die Probleme, die wir in München, in Berlin und in Frankfurt haben, andere sind als in Nairobi, in Mexiko und demnächst in Shanghai, geht es ja überall darum: Wie können wir Öffentlichkeit erzeugen?

Wie können wir Kommunikation erzeugen? Wie können wir soziale Verträglichkeit erzeugen? Bei aller Sehnsucht nach Mobilität – keiner wird sich mehr aufs Pferd setzen, sondern will eben 500 PS haben – ist das verträglich? Nur einfach zu sagen: Hier werden Fußgängerzonen ausgewiesen, dort das Schnellbahnsystem, wird nicht ausreichen.

Wie wird die AutoUniversität aussehen?

Die AutoUniversität gehört zu einer neuen Gebäudetypologie, die es so noch nicht gegeben hat, mit Strukturelementen, die eher Assoziationen an ein italienisches Bergdorf hervorrufen. Wir schaffen dort Räume, um andere Räume neu zu erschließen, um das Thema der Cluster-Bildung und der Future-Labs innerhalb einer neuen urbanen, komplexen Struktur zu ermöglichen.

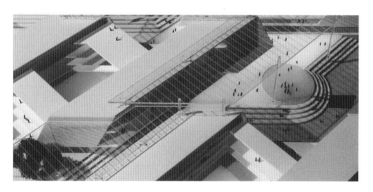

AutoUniversität Wolfsburg (vgl. a. Vorseite):
Neue Gebäudetypologie mit besonderer Clusterbildung.
AutoUniversität Wolfsburg (see also previous page):
New type of building with special cluster formation.

At AutoUniversität, the idea is to create knowledge space, space for new values and meanings.

The issue is also one of how far the mobility, or auto-mobility, topic has a future. What can we achieve with mobility?

That is addressed here. What is special about AutoUniversität are its "future labs". It is not only a university, but a combination of university and company, or rather business. The latter is represented by Motorola, Intel, Hewlett-Packard, or the Harvard Medical School. All these institutions will have future labs there, each with a staff of between five and ten people: developers, researchers or mavericks – lateral thinkers whose job is to decide where we can go from here. How will we have to see automobility ten years from now? How will physical or emotional mobility be created?

What does the "cars and cities" topic mean for you? Surely there is a considerable need for research here?

I can well imagine that the university will address this topic, and I would hope that one university department takes a very detailed, critical look at the future of the city and of urbanity, and at the connection between urbanity and mobility. Even though the problems we have in Munich, Berlin or Frankfurt are different from those in Nairobi, Mexico City or (soon) in Shanghai, the issue is the same everywhere: How can we generate public space? How can we generate communication? How can we generate social acceptability? How ever strong our desire for mobility may be – nobody is willing to get on a horse any more, but

wants to have 500 horsepower – is that acceptable? It will not be enough simply to say: "Here are your pedestrian precincts, there is your commuter rail network".

What will AutoUniversität look like?

AutoUniversität will belong to a new type of building, one that has never been seen, with structural elements that conjure up associations of an Italian mountain village more than anything else. We will create space there in order to open up other spaces, to allow the topic of cluster-building and future labs to develop within a new, urban, complex structure.

wo die autos herkommen:
fabrik und manufaktur

where the cars come from:
plants and manufactory

porsche in sachsen: kundenzentrum im werk leipzig

Porsche ist ein kleiner, aber dafür ausgesprochen prosperierender Autohersteller, er weitet seine sportive Kollektion aus und bietet mit dem Cayenne einen Geländewagen im Porschekleid an. Für eine jährliche Produktion von etwa 25.000 Exemplaren entstand in Leipzig zum ersten Mal außerhalb des schwäbischen Zuffenhausen ein Montagewerk, geschmückt mit Teststrecke und dem mittlerweile obligatorischen Kundenzentrum. Die Hamburger Architekten von Gerkan, Marg und Partner wählten eine signifikante Form aus: die kreisförmige Kanzel. Ein gigantischer Kreisel, wie man ihn als Kinderspielzeug kennt und er hier symbolisch an den ungehemmten Spieltrieb des Porschefahrers appellieren darf. Nur wenige architektonische Kunstgriffe reichen aus und sorgen für Markanz: Einmal steht der Kreisel als Blickfang unübersehbar in der Haupterschließungsachse, zum Zweiten wurden matt glänzende Metallpaneele für die Fassadenverkleidung verwendet, die auf das übliche Blechkleid der Autos anspielen. Im Übrigen gilt: Schreib drauf, wer du bist, und das so groß es geht: Architektur als Reklametafel!

Bauzeit _____ 2001–2002
Architekten _____ von Gerkan, Marg und Partner, Hamburg

porsche in saxony: customer centre at the leipzig plant

A small but extremely prosperous car manufacturer, Porsche is extending its range of sports cars, and has now added the Cayenne to its range, an off-road vehicle in Porsche clothing. Production of this model will be roughly 25,000 vehicles annually, and for this purpose Porsche has opened an assembly plant in Leipzig, the first time that production has been located outside the Stuttgart suburb of Zuffenhausen. The new plant is complete with test circuit and the now obligatory customer centre. Hamburg-based architects von Gerkan, Marg und Partner selected a suggestive form: a circular pulpit. The result is a huge "spinning top" reminiscent of the nursery, in this context appealing symbolically to the uninhibited playful urge of the Porsche driver. A few architectural tricks are all that is needed to bestow it with a distinctive air: for one thing, the top is an unmissable point en vue on the main connecting road, for another the matt sheet metal panels used as facing for the façade are an allusion to the panels commonly used on the exterior of cars. For the rest, the principle was "proclaim who you are in the largest letters possible": architecture as an advertising hoarding!

Constructed _____ 2001–2002
Architects _____ von Gerkan und Partner, Hamburg

Das Kundenzentrum bietet Raum für Ausstellungsflächen, Restaurant, Auditorien, den Streckenleitstand für die benachbarte Teststrecke, Shop und die Fahrzeugabholung (von oben nach unten).
Inside the customer centre there is room for display areas, a restaurant, auditoria, the control centre for the neighbouring test circuit, a shop and the handover area for new cars (from top to bottom).

form follows flow: die neue skodafabrik, mladá boleslav

Mit der Möglichkeit, ein Automobilunternehmen im ehemaligen Ostblock zu übernehmen, ergab sich für VW die Chance, mit einem Neubau heutige Erkenntnisse in der Autofertigung zu optimieren und Neues hinzuzufügen. Das Codewort bei VW dazu heißt „fraktal" – ein Begriff, der wohl ursprünglich aus der Mathematik stammt und Muster „von hoher Komplexität, wie sie in der Natur vorkommen", beschreibt (Skoda-Werksbroschüre). Wie beim menschlichen Stoffwechsel beispielsweise, der durch linear gedachte Abläufe nicht steuerbar wäre.

In einem modernen Unternehmen kann man von solchen Denkmodellen profitieren, indem man Selbstorganisation und Selbstoptimierung kleineren Regelkreisen und selbstständigen Arbeitszyklen überlässt – so weit verkürzt und fragmentarisch dargestellt.

„Die fraktale Fabrik" besteht im Wesentlichen aus dem Montageband und den „Fraktalen", die an das Montageband liefern. „Die Fraktale sind überschaubare Vormontagen oder Systeme, die geschlossene Qualitäts-Regelkreise darstellen" (Skoda).

An die Architektur werden dadurch neue Aufgaben gestellt, das heißt: „Form follows Flow"! Die Architekten des Büros Henn konnten auf umfangreiche Erfahrungen zurückgreifen, wo es um eine „Dynamik" in der Statik geht, also darum, Flexibilität in den Konstruktionen für entsprechende Fertigungsprozesse zu entwerfen. Außerdem wurde hier das Thema Kommunikation zum bestimmenden Entwurfselement, das heißt, die optimale und grenzenlose Förderung von Kommunikation zwischen allen, die das Auto fertigen. Zur Kommunikation gehört die Transparenz. Und so wird deutlich, warum ein Glas-Stahl-Outfit geboten scheint.

Hinweis: Dieser neue Werksteil, die Montagehalle, ist nur eine kleine Ergänzung bestehender Werksanlagen in Mladá Boleslav. >

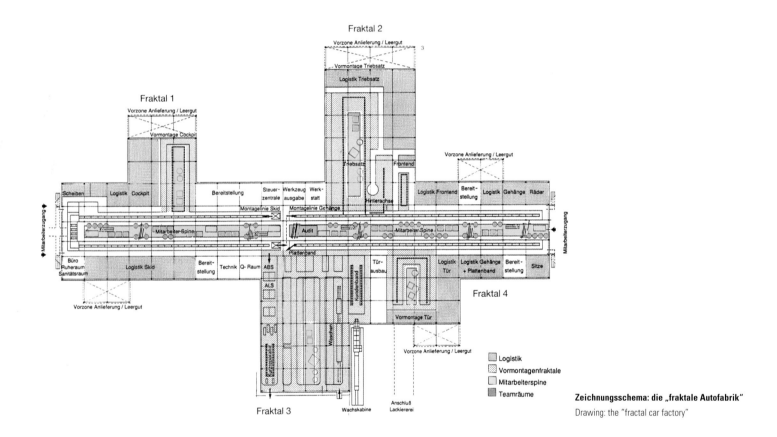

Zeichnungsschema: die „fraktale Autofabrik"
Drawing: the "fractal car factory"

form follows flow: the new skoda factory, mladá boleslav

For VW, the possibility of acquiring a car manufacturing company in the former eastern bloc was an opportunity to create a new building including state-of-the-art car manufacturing techniques and new elements. The magic word linking all these efforts at VW was "fractal" – a term which has its origins in mathematics and which, to paraphrase the Skoda factory brochure, describes patterns of high complexity that exist in nature. One example of this is the human metabolism, which could not be controlled using a process conceived in linear terms.

To put it very briefly and fragmentarily, such models of thought can be exploited in a modern company by leaving self-organization and self-optimization to small self-regulating groups and independent work cycles.

The "fractory" essentially comprises the assembly band and the "fractals" feeding into that band. In Skoda's words, "the fractals are manageable pre-assembly elements or systems forming closed, self-regulating quality groups". This sets new tasks for architecture, the byword being "form follows flow". The architects from the Henn bureau were able to draw on their extensive experience of instances in which the theory of structures is concerned with "dynamics", and thus, in the structures to house such manufacturing processes, to come up with designs in which flexibility plays a part. At the same time, the communication theme became the decisive design element. In other words, the best possible, boundaryless promotion of communication between all those manufacturing the car. Transparency is an integral part of communication. And this explains why a glass and steel design would appear to be appropriate.

It should not be forgotten that this new part of the factory, the assembly shop, is only a small addition to the existing works complex in Mladá Boleslav. >

Kopf- und Eingangsgebäude von außen und innen gesehen Main building and entrance building seen from the outside and the inside

Zentrale Zone der Montagehalle
Central zone of the assembly shop

Ausschnitte aus der Skoda-Montage
Details of Skoda assembly shop

In der zentralen Zone der Montagehalle liegen nicht nur die Montagelinien der Fertigung, sondern auch die Aufenthalts- und Logistikbereiche der Mitarbeiter, hier werden die Vorprodukte zugeführt. Architektonisch reizt es natürlich, die funktionellen Vorgaben, in diesem Fall also die eigentliche Fertigung, symbolisch zu überhöhen und zu betonen. Es geschieht hier durch hoch gesetzte Dächer für eine zusätzliche Belichtung, fast wie in einer Basilika. Und noch etwas ist prototypisch für diese Autofabrik der Zukunft: Hier sind die Büros in die Fertigungsbereiche integriert – white collar meets blue collar.

Bauzeit	1995–1996
Maße	Modulfelder im Fraktalbereich: 18 x 18 m
	Lichte Höhe: 15 m, Mittelteil: 41 x 216 m
Bebaute Fläche	32.000 qm
Konstruktion	Stahlkonstruktion mit Baumstützen,
	Stützen: Stahlbeton, Dachkonstruktion: Stahl
Architekten	Henn Architekten, München

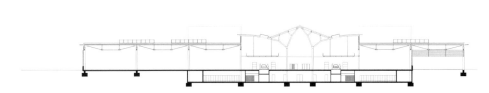 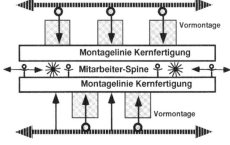

Zeichnungen: Querschnitt (links), Montageschema mit zentralem Fertigungsbereich (rechts)
Drawings: section (left), assembly plan with spine (right)

Not only assembly lines for production are situated in the central zone of the assembly shop, but also employees' recreational areas and logistics areas, and fabricated products are fed into the spine here. Architecturally, of course, it is tempting to exaggerate symbolically and emphasize the actual manufacturing process. Here, this is done by raising the roof for additional lighting, almost like in a basilica. And there is something else that is prototypical for this car factory of the future: here, offices have been integrated in the manufacturing areas – white collar meets blue collar.

Constructed	1995–1996
Dimensions	Modular squares in the fractal area: 18 x 18 m
	Clear height: 15 m, Middle section: 41 x 216 m
Coverage	32,000 m²
Construction	Steel structure with tree-shaped supports, reinforced concrete
	supports, steel roof structure
Architects	Henn Architekten, Munich

**elegant aufgefächert:
design center innerhalb des mtc, sindelfingen**

Mercedes-Benz hat alle Kräfte, die an der Neukonzeption der Fahrzeuge beteiligt sind, im neuen MTC (Mercedes Technology Center) gebündelt. Dies sollte symbolisch unter einem Dach geschehen, in Wirklichkeit liegen zumindest alle Bereiche nahe beieinander. Für den Teilbereich „Design Center" wählte der mit dem Bau speziell beauftragte Architekt Renzo Piano das Dach als große schützende Form und schlägt es geschickt als Fächer auf.

Renzo Piano baut so einiges für den Konzern mit dem Stern. Er ist aber nicht das, was man einen „CI-Architekten" nennen möchte. Zusammen mit seinem „Building Workshop" ist er ein technisch-konstruktiv denkender Mensch, der diesen Ansatz mit funkelnder Ästhetik kreuzt. Die Lage des Design Center in einer großen Kurve gleich als Auftakt des gigantischen Sindelfinger Werks hat zu einer dynamisch schwingenden Staffel von riesigen Sheddächern geführt, die auf diese Weise gutes Seiten- und Oberlicht

in die Design- und Denkschmiede lassen. Renzo Pianos architektonisch-ingenieursmäßiges Feuerwerk gipfelt außen in einem Merkzeichen mit sichtbaren Stahlrohrkonstruktionen, Rohrstreben und Zugankern – notwendige, aber auffällige Helfer, die die jeweils um neun Grad gegeneinander verdrehten Dachscheiben tragen und sichern. Es ist ein lustvoller, Aufmerksamkeit heischender Bau, besonders mit seinem gläsernen Gesicht zur öffentlichen Straße hin.

Leider ist das Design Center abgeschottet wie ein Hochsicherheitstrakt. Gesichert auch vor lästigen Blicken ist die Kreativität der Designer und Ingenieure, und deswegen ist konsequenterweise dieser Gebäudeteil von dichten Anpflanzungen verborgen und uneinsehbar. Keine Chance für fernöstliche Werksspionage. >

elegantly fanned out:
design center in mtc, sindelfingen

Mercedes-Benz has pooled together all those employees involved in the initial design of vehicles in its new MTC (Mercedes Technology Center). Symbolically at least, all this is meant to be contained under one roof, in reality all the departments are at least in the vicinity of each other. The architect engaged to build the "Design Center" section, Renzo Piano, chose the theme of the roof as a large protective form, skilfully opening it out like a fan.

Renzo Piano has built a fair number of buildings for the automotive group with the star logo; but he is not what would normally be called a "CI architect". In the work he does together with his "Building Workshop", he is someone who thinks in terms of technology and structures, and who crosses this approach with aesthetic brilliance. The location of the Design Center in a long curve close to the entrance of the huge Sindelfingen plant has given rise to a dynamic formation

of huge sawtooth roofs, a solution which allows good light to enter this hotbed of ideas and design through fanlights and sidelights. On the exterior, Renzo Piano's architectural and engineering fireworks culminates in a landmark with visible tubular steel structures, tubular braces and tie rods – necessary but conspicuous helpers, which secure and take the load of the roof sections, each of which has been rotated at nine degrees to its preceding section. It is a zestful building that demands attention, especially with its glass front facing the public road.

Unfortunately, the Design Center is shielded from public access like a high-security wing. At the same time, however, the designers' and engineers' creativity is shielded from prying eyes, and so it only makes sense that dense vegetation has been planted to conceal it. Industrial espionage cannot get a look in. >

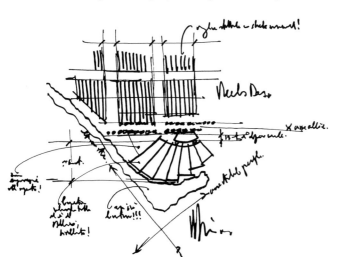

Am Anfang steht die Idee: Skizze des Architekten Renzo Piano.
It starts with an idea: sketch by the architect Renzo Piano.

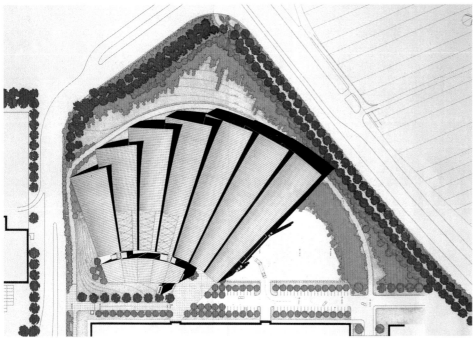

**Dynamisch gebogene Dachsegmente prägen
das Gesicht des Design Center.**
Dynamically curved roof sections give the Design
Center its distinctive appearance.

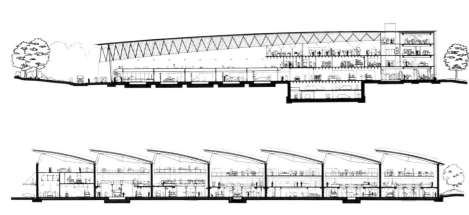

Längs- und Querschnitt Longitudinal and cross-section

Der aufgeschlagene Dach-Fächer aus der Luft gesehen.

A bird's eye view of the roof's fan-like form.

Weil nicht einmal den anderen Werksangehörigen der Zutritt gestattet ist, hier ein Blick hinein: Piano vergleicht den Grundriss mit einer ungewöhnlichen Hand mit sieben Fingern. In diesen „Fingern" arbeiten Designer, Modellbauer und andere jeweils für sich, aber mit Sichtkontakt. Front- und Dachfenster sorgen für einen angemessenen Anteil an Tageslicht – zur Inspektion der Autoneulinge.

Virtuoses Filetstück der Anlage ist sicher das dreidimensional ausgestaltete Dach. Renzo Piano dazu: „Die dreidimensionale Dachform wurde auf der Grundlage einer toroidalen Geometrie entwickelt („der Form eines Autoreifens nachempfunden", Anm. des Autors), und die doppelte Krümmung wird von einem ausgeklügelten System aus Stützen, Gelenken und Zugelementen getragen, das der Dachwölbung die Spannung verleiht. Das eigentliche Dach besteht aus einem Sandwich, also aus zwei Stahloberflächen, zwischen denen sich eine dicke Schicht aus wärmedämmendem Material befindet." Das „Haus der kreativen Köpfe", wie es der damalige Chefdesigner Bruno Sacco gern nannte, hat also eine kongeniale Form und Konstruktion. Es bukt sich an einen seriös einfachen Bürokopfbau an und bildet den Auftakt für das riesige MTC mit seinen ca. 6.000 Arbeitsplätzen.

Bauzeit	1994–1996
Konstruktion/Material	Spritzbeton, Aluminiumpaneele, Stahlkonstruktion
Architekten	Renzo Piano Building Workshop, Genua und Christoph Kohlbecker, Gaggenau

Dachträger unter transparentem Dachausschnitt
mit Lichtsteuerelementen (Präsentationshalle).
Rechts: Blick ins Allerheiligste – ohne Autos
Roof beam under a transparent roof section
with light control elements (presentation hall).
Right: A view of the inner sanctum – without any cars

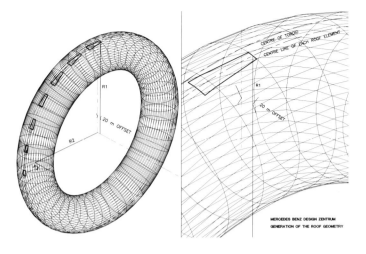

MERCEDES BENZ DESIGN ZENTRUM
GENERATION OF THE ROOF GEOMETRY

Entstehung des Dachs als Ausschnitt aus einem Toroid (ähnelt einem Autoreifen).
The roof is formed as a segment of a toroid (similar to a car tyre).

Because even plant employees are not allowed to enter the building, here is a look inside: Piano compares the layout with an unusual hand with seven fingers. Designers, modellers and others work in each of these "fingers" – kept separate from each other, but with visual contact. Lights in the façade and roof allow an adequate amount of daylight in – in which to check the function of new cars.

The pièce de résistance of the complex is without doubt the three-dimensional design of the roof. Renzo Piano comments: "The three-dimensional form of the roof was developed on the basis of a toroid geometry [author's note: i.e. loosely based on the form of a car tyre], and the double curvature is supported by a sophisticated system of supports, joints and tie elements which provides the

arched roof with the tension it needs. The roof itself is made up of a sandwich, i.e. two steel surfaces between which there is a thick layer of heat insulating material." In other words, this "house of creative minds", as former chief designer Bruno Sacco liked to call it, has a congenial form and design. It nestles up against a respectable, simple main office building and is the prelude, as it were, to the huge MTC with its roughly 6,000 workplaces.

Constructed	1994–1996
Construction/materials	Shotcrete, aluminium panels, steel structure
Architects	Renzo Piano Building Workshop, Genoa and Christoph Kohlbecker, Gaggenau

„Autofabrik" oder „höfisches Lusthaus" (*Die Zeit*)? In jedem Fall steht die Gläserne Manufaktur in
Dresden für die Rückkehr der Industriearbeit in die City, und sie ist eine Bereicherung für die Barockstadt.
Car factory or "courtly pleasure palace" (*Zeit*)? At all events, the glass factory in Dresden stands for the return of
industrial work to the city centre, and it belongs to and enhances that Baroque city.

sächsische traumfabrik: gläserne manufaktur, dresden

Eine Autofabrik im Großen Garten, der barocken Urzelle Dresdens? Ja – dort steht ein gläsernes Konvolut aus wuchtigen Türmen und Kuben und veranstaltet abends das „große Leuchten" (*Süddeutsche Zeitung*). Die Besucher erreichen beim Zutritt zunächst eine haushohe Lobby, deren metallisches Blinken an Cape Canaveral erinnert und in eines dieser futuristischen Hotels in Fernost oder Dubai führen könnte. Doch die Aufzüge und Rolltreppen transportieren die Besucher dann vor den gläsernen Vorhang einer Bühne, deren lautlose Inszenierung zunächst kein Thema hat, weil die Protagonisten sich nur sehr langsam bewegen und die zahllosen Fahrzeuge dort nur zu bewundern, bisweilen zu streicheln scheinen. Tatsächlich wird hier die komplette Montage einer Edelkarosse des VW-Konzerns sichtbar als öffentliches Theaterstück inszeniert. Der Name: Phaeton, frei übersetzt Feuerwagen oder „der Leuchtende". >

saxon dream factory: factory of glass, dresden

A car factory in the Grosser Garten park, the germ-cell of Dresden's Baroque cityscape? Yes, there it sits, a glass gathering of massive towers and cubes which, the evenings, stages a "great illumination", to quote the *Süddeutsche Zeitung*. On entering the building, visitors first find themselves in a foyer as high as the building itself, a metallic environment of flashing lights reminiscent of Cape Canaveral, which might be the anteroom to one of those futuristic hotels you find in the Far East or Dubai. But the lifts and escalators then transport visitors along the glass curtain of a stage, on which a performance is being acted out silently and, at first glance, without any plot, as its protagonists move only very slowly and appear solely to be admiring the many limousines there, if not stroking them. What is in fact happening here is the complete assembly of a luxury VW limousine, conspicuously staged as a public performance. Its name is Phaeton, which, if translated freely, means wagon of fire or the "radiant one". >

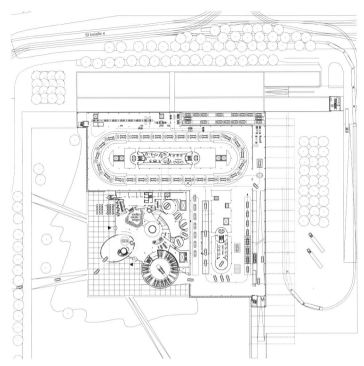

Grundriss Plan

Die Collage der Baukörper ist ein Ballett der geometrischen Formen, hat allerdings auch die Aufgabe, dem großen Hauptglaskörper die Wucht zu nehmen und den „städtischen Maßstab" zu halten.

The collage of buildings is a ballet of geometric shapes, but also has the task of diminishing the weightiness of the huge main glass building and of maintaining an "urban scale".

„Ein Auto ist etwas Bewegliches, also sind Gebäude dafür gut, die Bewegung, also Abläufe sichtbar zu machen". Aus diesem Statement des Architekten Gunter Henn ergibt sich das Leitkonzept von Transparenz und Offenheit. Die Gläserne Manufaktur verbirgt nichts. Administration, Manufaktur und die Öffentlichkeit zusammen im Glashaus. Wer als Besucher oder Abholer auf der Empore steht, dem bleibt nichts, wirklich nichts verborgen: auch nicht die imposanten Motorblöcke, wie sie im mächtigen Glasfahrstuhl nebst Chassis nach oben gefahren werden.

Glas auch außen – kein steinerner Fabrikauftritt mehr, sondern Fassaden, in denen sich das Blau des Dresdner Himmels und das Grün des Parks spiegeln. Die Industrie stößt hier keine gefährlichen Emissionen mehr aus. Große Lastwagen und Sattelschlepper für die Logistik, die tiefe Reifenspuren im Park hinterlassen, sucht man so gut wie vergebens. Es gibt eine elegante „CarGo Tram", eine Straßenbahn, die die Montageteile von einem Außenlager herbeischafft und tief hinein in den Bauch der gläsernen Fabrik fährt. >

"A car is a moving object, and so buildings are good for making movement, i.e. processes, visible". This statement by the architect Gunter Henn explains the guiding concept of transparency and openness. The glass factory does not conceal anything. Administration, manufacture and the general public, together in an all-glass building. From the gallery, nothing, truly nothing, is hidden from the visitor or customer who has come to pick up his car: not even the impressive engine blocks, as they are brought up in the huge glass lift, along with chassis assemblies.

Glass is on the exterior too – the factory no longer presents itself in stone, but with green-blue facades, in which the blue of the sky over Dresden and the green of the park are reflected. Here, industry no longer belches out dangerous fumes. A search for huge lorries and articulated trucks delivering parts and leaving deep ruts in the green of the park will be more or less in vain: instead, there is an elegant "CarGo Tram", a tram which collects the parts for assembly from an external warehouse and transports them deep into the bowels of the glass factory. >

Oben: Zutritt zum gläsernen Foyer, das mit Restaurant und Event-Bereich zu einem gelungenen „architektonischen Abenteuerspielplatz" (*Bauwelt*) geworden ist (unten).
Top: Entrance to the glass foyer. With its restaurant and event space, it is, in the words of *Bauwelt* magazine, a fine example of an "architectural adventure playground" (below).

Je weiter der Besucher nach oben in Richtung Autofertigung vordringt, desto überraschender
werden die Formen und Details. Oben: Foyer. Unten: Blick zurück und nach unten.

The closer the visitor comes to actual car manufacture, the more surprising the forms and details
become. Top: foyer. Below: looking back and down.

**Transparenz ist erste Pflicht in der Gläsernen Manufaktur. Verwaltung und
Fertigung sind nicht mehr getrennt (siehe auch vorherige Seiten und rechts).**
At the glass factory, transparency is paramount. Administration and manufacture
are no longer separate (see also previous pages and right).

**Rechts oben: Ein Ambiente wie in einem Tanzsaal oder einer Agentur:
Montageband auf Parkettboden.**
Top right: An atmosphere like in a ballroom or an agency: assembly band
on parquet flooring.

Die Fertigung in Dresden beginnt nicht mit der ersten Schraube. Chassis, Motorblock und die fertig lackierte Karosserie, die dann hier ausgerüstet, also vor allem mit Unmengen an Elektronik gefüllt wird, werden zusammengefügt, „verheiratet", wie es im Branchenjargon heißt. 36 Stunden dauert es, bis ein Phaeton endgefertigt ist. Es wird im 24-Stunden-Betrieb gearbeitet, in einer Umgebung, die man als den Industriearbeitsplatz der Zukunft bezeichnen muss und die die optischen Qualitäten einer Werbeagentur besitzt: Frühstücksplätze auf Designerstühlen mit Blick in den Großen Garten, Parkettfußboden wie im Restaurant, auch für das Montageband. Kein Beton, kein Estrich, denn hier fahren keine lauten Gabelstapler herum, sondern elektronisch gesteuerte kleine Boxen, mit Autoteilen und Werkzeugen. Es gibt eine durchgehende „Henn"-Handschrift, die erkennbar bleibt, eine „Durchgängigkeit in der gestalterischen Haltung" – in Glas und in Metall. Das ist auch von der Autostadt und dem Skodawerk her bekannt.

Nicht mehr erkennbar ist, was Werkzeug, was Konstruktion ist: „Es gibt keine Schnittstelle zwischen Architektur und Einrichtung", die Skips (diese neue verdeckte Art des Förderbands) und Aufzüge sind integrale Bestandteile des Projekts. Alles Design stammt von diesem Münchner Architekten. Zu Henns Auftrag gehört ausdrücklich dessen persönlicher Ästhetik-Check alle paar Wochen, und inzwischen auch das Blumenarrangement in der Lounge, wo Kunden etwas trinken können, bevor sie ihr Auto abholen.

Bauzeit	2000–2002
Maße	Länge: 147 m, Breite: 147 m
Nutzfläche	Produktionsbereich: 55.000 qm
Glasfassaden	26.700 qm
Parkett	19.000 qm (Ahorn + Eiche)
Arbeitsplätze	ca. 800
Architekten	Henn Architekten, München

In Dresden, production does not start right from the first screw. The chassis, engine block and ready painted body, which is then equipped here, i.e. filled with an endless array of electronic componentry, are brought together or "married", to use the industrial jargon. After 36 hours, the assembly of a Phaeton is complete. Work is done round the clock, in surroundings that have to be described as the industrial workplace of the future, and seem more like an advertising agency: breakfast places on designer chairs, looking onto the Grosser Garten, parquet flooring like in a restaurant (even for the assembly line). No concrete, no screed, for there are no noisy fork-lift trucks driving around here, but electronically controlled little boxes containing car parts and tools. It bears the unmistakeable Henn trademark, an uncompromising design principle in glass and metal. This is the same principle that was manifested at Autostadt and the Skoda factory.

It is no longer apparent what is utility and what is design: "There is no interface between architecture and fittings", the skips (as the new concealed type of conveyor belt is called) and lifts are integral parts of the project. The entire design is the work of this one Munich architect. Henn's personal aesthetic sense is also an explicit part of his job – he checks how things look every few weeks, right down to the flower arrangements in the lounge where customers take a drink before picking up their car.

Constructed	2000–2002
Dimensions	Length: 147 m, width: 147 m
Usable space	Production area: 55.000 m²
Glass façades	26,700 m²
Parquet flooring	19,000 m² (maple and oak)
Workplaces	approx. 800
Architects	Henn Architekten, Munich

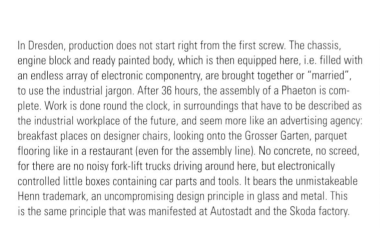

Haben Sie sich schon öfter mit diesem Thema, mit „MotorTecture", befasst?

Ja, zum Beispiel haben wir zusammen mit den Architekten Kauffmann und Theilig für die Markteinführung der A-Klasse für DaimlerChrysler eine Wanderausstellung konzipiert und das architektonische Konzept mit unseren ingenieursmäßigen Aufgaben verwoben. Die A-Klasse musste innerhalb eines Sommers in mehr als 30 Städten präsentiert werden, das heißt, es war das große logistische Problem der schnellen Montage und Demontage zu lösen. Das ist uns gelungen. Wir haben mit Christoph Ingenhoven einen Messe-Pavillon für Audi konzipiert (s. a. folgende Seiten) und gehören zum Team für das Mercedes-Benz Museum (s. S. 172 ff).

Inwieweit ist es befruchtend, sich vom Kollegen Autobauer etwas abzuschauen? Was können Autobauer, was Bauingenieure und Architekten nicht können?

Eine perfekte Standardisierung der Bauteile auf einem durchgehend hohen unverwundbar festgeschriebenen Qualitätsniveau. Sie haben eine Befähigung zur absoluten Entmaterialisierung und Leichtigkeit – und können dünner, immer dünner konstruieren, damit immer weniger Benzin verbraucht wird. Im Bauwesen ist es auch notwendig – bei Brücken usw. – immer mehr Material zu sparen, und es ist eine ökologische Notwendigkeit, so vorzugehen und das Gleiche mit einer Tonne Aluminium zu erreichen, wenn man früher zehn einsetzen musste.

Sie haben für sich ein Glashaus gebaut, das als Experimentierprojekt für das Bauen gilt, haben Sie dort auch solche Erfahrungen gemacht?

Mein Haus wiegt ein Sechstel eines normalen Hauses, warum soll es schwerer sein, wenn es leichter geht. Unser Credo heißt, Dinge leicht und verständlich zu machen, mit hoher Präzision und Vorfertigung auf der Baustelle fertigzustellen und nicht nach herkömmlicher Gummistiefeltheorie im Schlamm zu stehen, sondern in einem fast schon ästhetischen Prozess so zu verbinden, dass man nach Ende des Gebrauchs des Bauwerks, ob nun nach drei Tagen oder tausend Jahren, es genauso schnell und einfach wieder auseinandernehmen kann. Und für diese Einwerkstofftechnologie habe ich viel bei der Automobilindustrie gelernt, weil es schon seit 15 Jahren entsprechende Konzepte gibt. Mich beeindruckt, wie Autobauer ein Detail, also z.B. einen Türknopf entwickeln, d.h. bezeichnen, beziffern und entwerfen. Wie die Montage und Demontage geplant wird.

Können die Autobauer von den Bauleuten lernen?

Ich könnte nicht sofort sagen, was. Das bedeutet aber nicht, dass ich die Frage verneinend beantworte. Es ist so, dass die unheimliche Anhäufung von Wissen nicht mehr von einzelnen Gruppen, sondern verknüpft und interdisziplinär benutzt wird und entsprechende Synergien schafft. Ich veranstalte jährlich an der Universität Stuttgart ein Symposium, wo Biologen, Chemiker, Autobauer, Glashersteller oder Architekten über ihre neuesten Forschungstrends berichten. Zunächst scheint das zusammenhangslos zu sein, nach zwei Tagen erkennen dann alle Teilnehmer, wie sensationell gut sich die Erkenntnisse verknüpfen lassen!

→ **interview** interview

Werner Sobek ist Inhaber des Büros Werner Sobek Ingenieure in Stuttgart-Degerloch und hat eine Professur an der Universität Stuttgart inne. Er zählt zu den profiliertesten deutschen Bauingenieuren und hat u.a. am Audi-Messepavillon zusammen mit Christoph Ingenhoven gearbeitet. Und er gehört auch zum Team, das das neue Mercedes-Benz Museum bauen wird.

Werner Sobek is the owner of the Werner Sobek Ingenieure bureau in Stuttgart, and holds a chair in Engineering at the University of Stuttgart. He is one of the most outstanding structural engineers in Germany, his work having included collaboration with Christoph Ingenhoven on Audi's trade fair pavilion. Furthermore, he is part of the team that will build the new Mercedes-Benz Museum.

Have you often been concerned with the "MotorTecture" topic?

Yes. For example, together with the Kauffmann and Theilig architects' office, we designed a road show for the market launch of DaimlerChrysler's A class, weaving this architectural concept into our engineering task. In the course of one summer, the A class had to be presented in more than 30 cities, which meant that we had to solve a substantial logistical problem of rapid assembly and disassembly. We succeeded in this. We worked together with Christoph Ingenhoven on Audi's trade fair pavilion (see also following pages) and we are part of the team working on the new Mercedes-Benz Museum (see p. 172 ff).

How fruitful is it to look over automotive engineers' shoulders? What can they do that structural engineers and architects cannot?

Perfect standardization of components on a quality level that is consistently high and irrevocably fixed. They are skilled in absolute dematerialization and lightness – they are able to design ever thinner, so that less petrol is consumed. Using construction materials sparingly is also necessary in building – for bridges, etc. – and there is an ecological necessity to do so, achieving the same effect with one ton of aluminium that used to be achieved with ten.

You have built yourself a house of glass, one which is regarded as an interesting building project. Was this something you experienced here, too?

My house weighs one-sixth of a normal house – why make it heavier, if lighter is also possible? Our motto is to make things light and understandable, to complete the building process on site with a high level of precision and pre-fabrication, not standing around in the mud as with the traditional gumboot theory, but combining them in a process that is almost aesthetic. The result should be that once the building no longer needs to be used, be it after three days or a thousand years, dismantling it should be just as quick and easy to do. This single-material theory owes a lot to the automotive industry, where such ideas have been around for 15 years now. I'm impressed by the way an automotive engineer develops (i.e. describes, specifies and designs) an individual feature such as a door knob. Also by the way that assembly and disassembly is planned.

Can car producers learn from builders?

I couldn't immediately say what they can learn. That's not to say that my answer is "no". The point is that the incredible wealth of knowledge is no longer used by individual groups, but in interlinked, interdisciplinary ways, resulting in synergy. Every year I organize a symposium at the University of Stuttgart, at which biologists, chemists, car producers, glass manufacturers or architects report on the latest research trends in their disciplines. At first it all seems unrelated, but then after two days all the delegates realize how fantastically well this knowledge can be combined.

**corporate architecture:
showroom und messepavillon**

corporate architecture:
showrooms and fair stands

der gläserne autoreifen: messestand für die audi ag

„Der Evolutionsprozess der Lebewesen und Pflanzen kann der bestimmende Ratgeber zukünftiger Generationen von Ingenieuren und Architekten sein", schreibt der Architekt Christoph Ingenhoven. Diese Erkenntnis hat er auch beim Auftritt von Audi auf Automessen zwischen Frankfurt und Tokio angewendet. Es ist ein Schnittstellenprojekt geworden, nicht nur zwischen Auto und Architektur, sondern auch zwischen biomorphem Design und Medieninstallation, in Arbeitsteilung von Architekten der Gruppe Ingenhoven Overdiek und Partner und den Designern bzw. Marketingleuten von KMS aus München.

Das Ergebnis gehört zur Gattung der „Fliegenden Bauten", entstand ohne Beton und Stein. Die Gestalter haben einen Messestand erdacht, der an einen riesigen gläsernen Autoreifen erinnert. So kann man zunächst durchaus glauben, der Pavillon wäre der neuen Allianz der Blobs zuzurechnen,

die aus Amerika die Architekturszene überschwemmt wie Nike-Schuhe die Laufstrecken der Welt. Es ist aber zuerst einmal ein Konstrukt aus Edelstahlrohren, Glasscheiben und Seilnetzen. Ein „loop" genanntes Gebilde, das Korpus und – für den Messebetrieb so wichtig – Projektionsfläche in einem ist. Ein temporärer, ein demontabler Bau. Selbstverständlich auf den Spuren geschätzter Vorbilder wie Frei Otto (der Konstrukteur des Münchner Olympiadachs 1972) und unter Mitarbeit eines der profiliertesten Ingenieure von heute, Werner Sobek aus Stuttgart. Es ist ein Grenzgang der Architekten.

Baujahr	1999
Konstruktion	Edelstahl, Glas, Seilnetze
Autoren	Ingenhoven Overdiek und Partner, Düsseldorf,
	KMS Team, München, Werner Sobek Ingenieure, Stuttgart

the glass car-tyre: trade fair stand for audi ag

The architect Christoph Ingenhoven writes: "The evolution of living organisms and plants can show the way for future generations of engineers and architects". He has applied this insight to Audi's exhibitions at car shows from Frankfurt to Tokyo. It has become an interface project, not only between car and architecture, but also between biomorphic design and media installation, the work being shared between architects from the Ingenhoven Overdiek und Partner group and the designers and marketing people from KMS in Munich.

The result was a product on the "temporary buildings" theme, without concrete or brick. The designers have thought up a trade fair stand that is reminiscent of a huge glass car-tyre. At first, therefore, one might well think that the pavilion has to be seen as part of the new alliance of blobs that is swamping the architectural scene from America, like Nike shoes on the world's racetracks. First

of all, though, it is a structure comprising tubular stainless steel, glass panes and rope meshing. This structure, called a "loop", is a combination of carcass and (extremely important for the trade fair business) projection surface. A temporary, knock-down building. It goes without saying that it follows in the footsteps of revered role models such as Frei Otto (who designed the roof of the 1972 Munich Olympic stadium), and was built with the involvement of Werner Sobek from Stuttgart, one of today's most outstanding engineers. For architects, it is a borderline experience.

Constructed	1999
Structure	Stainless steel, glass, rope meshing
Authors	Ingenhoven Overdiek und Partner, Dusseldorf,
	KMS Team, Munich, Werner Sobek Ingenieure, Stuttgart

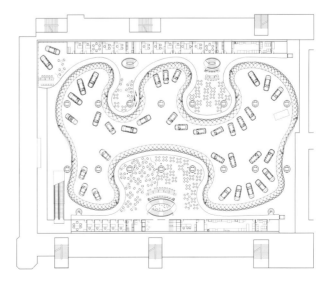

Oben: Autopräsentation im Loop. Links: Die unterschiedlichen Ausstellungsbereiche sind durch Torbögen zum einheitlichen Raum verbunden. Rechts: Grundriss

Top: Car presentation in the "loop". Left: The various exhibition areas are linked via archways with the shared room. Right: ground plan

Das Vorbild wird deutlich: ein Autoreifen

What inspired it is clear: a car-tyre

rund und farbig – die autoboutique:
saab city center, hamburg und münchen

Uhren- und Schmuckgeschäfte gehören in die Schokoladenlagen der Cities, Herrenausstatter und Damenboutiquen mit Edelmarken genauso. Und Showrooms von Edelkarossen? Ein Trend der letzten Jahre sorgt für die Rückkehr von Autohäusern in die Innenstädte, und Saab gehört in Deutschland zu den Trendsettern. Die neuen City Center setzen auf Premiumlagen in ausgesuchten Passagen und Geschäftsstraßen. Präsentiert wird ein Lebensgefühl, eine Philosophie. Die Autos werden weiterhin in den Saab Partnerbetrieben vertrieben, doch wird das gesamte Leistungsangebot auch über die City Center vermittelt. Flachbildschirme als Botschafter der virtuellen Saab Welt stehen zur interaktiven Information bereit, selbstverständlich auch die obli-

gatorische Kollektion der Saab Accessoires. Je nach Größe des City Centers werden zwischen zwei und vier Autos ausgestellt. Die Innenarchitektur des schwedischen Architekten Wilhelmson profitiert vom heutigen hohen technischen Fertigungsstand der Glasindustrie und der EDV. Gebogenes Opalglas bildet die lichte Höhle für sportive Automobile. Ein ausgefeiltes EDV-Programm sorgt für wechselnde und verkaufsfördernde Lichtklimata.

Baujahr ———————— 2002
Architekt ———————— Anders Wilhelmson, Stockholm

round and colourful – the car boutique:
saab city center, hamburg and munich

Jewellers' shops belong in the prime city centre locations, as do men's outfitters and women's boutiques selling up-market brands. What about the showrooms of top-of-the-range cars? In recent years, there has been a tendency for car showrooms to return to city centres, and Saab has been one of the trendsetters in Germany. The new centres at the heart of cities are interested in premium locations in select malls and shopping streets. What is presented is a feeling of life, a philosophy. The cars are still sold in the Saab partner operations, but an idea of the whole range of services offered is also conveyed via its City Center. Flat screens, ambassadors of the virtual Saab world, are on hand to provide interac-

tive information, and of course there is the obligatory collection of Saab accessories. Depending on the size of the City Center, between two and four cars are on display. The interior design of Anders Wilhelmson, the Swedish architect, stands out by virtue of the high technical standards in modern glass production and in IT. Curved opal glass forms the light cave for sporty cars. A sophisticated IT program provides changing and sales-promoting light moods.

Constructed ———————— 2002
Architect ———————— Anders Wilhelmson, Stockholm

Saab City Center in Hamburg: beschwingte Architektur von Auto und Shop.
Saab City Center in Hamburg: dynamic architecture in the cars and the shop.

Saab City Center in München: Hinter historischen Fassaden Lichtprogramme, frech und überraschend wie in der Diskothek.
Saab City Center in Munich: behind historical façades playful and surprising light shows, like in a disco.

300-mal luxus: mercedes-welt, berlin

Berlin steht, was seinen Westen betrifft, schon seit geraumer Zeit unter dem Stern, der seit knapp 40 Jahren das Europa-Center krönt. Mit der Hauptstadtfunktion und ihren notwendigen Fahrzeugparks für Regierung und Diplomatisches Corps ist Berlin nun ein sehr wichtiger Markt für Prime Brands geworden, vielleicht der wichtigste in Deutschland, und so natürlich der Standort für den Piloten einer „Mercedes-Welt". Die „fließt" nun gut 150 m das Charlottenburger Salzufer am Landwehrkanal entlang hinter grün glänzendem Glas und unter dem Segel einer Schiffsmetapher, weil der spitze Hausbug zum Markenzeichen geworden ist.

Für Mercedes-Benz ist diese Mercedes-Welt so etwas wie das Abrücken vom Ingenieurs- und Technik-Image eines typischen Autohauses und der kongeniale Auftritt zum weltweit erfolgreichen „highway outlet", den autarken Marken-Stationen, die gut erreichbar an den Schnell- und Ausfallstraßen liegen und viel, sehr viel Platz verschlingen.

Das Mercedes Center ist „ein neues Vertriebsformat", eines mit Komplettanspruch, denn hier stehen bis zu 300 Modelle aller Couleur aus dem Haus mit dem Stern. Ein Service Center gibt es zwar auch, vor allem ist es aber ein Event Center: Willkommen zur Formel-1-Übertragung und zur Fußballweltmeisterschaft.

300times luxury: mercedes-welt, berlin

As far as the western part of the city is concerned, Berlin has been under a lucky star for some time – the star that has been atop the Europa Center for almost 40 years. Now that it is the capital, with the need for vehicle fleets for the government and the diplomatic corps, Berlin has become an extremely important market for prime brands, if not the most important in Germany. Not surprisingly, therefore, it is the location for the pilot "Mercedes-Welt" (Mercedes world). Along Charlottenburg's Salzufer, on the banks of the Landwehr Canal, this building "flows" for 150 metres behind shimmering green glass and under the sail of a ship metaphor, because the building's pointed bow has become a trademark.

For Mercedes-Benz, this Mercedes-Welt is in some ways a move away from the engineering and technological image of a typical car showroom, and a congenial launch of the highway outlet that is successful all over the world – the self-contained brand station that is within easy access of expressways and arterial roads and devours a great deal of space.

The Mercedes Center is a "new sales format" with a claim to completeness, containing as many as 300 models of all kinds from the company with the star logo. While a service centre also exists, it is primarily an event centre: welcome to televised Formula One and the World Cup.

Fassadenausschnitt: High-Tech-Architektur für High-Tech-Karossen.
View of the façade: high-tech architecture for high-tech limousines.

Die dramatisch zugespitzte Glasfront (Schiffsbug!)
korrespondiert mit der Uferlinie am Landwehrkanal.
The dramatically pointed glass façade (ship's bow)
corresponds with the course of the Landwehr Canal bank.

Ganz im Gegensatz zum Sindelfinger Design Center (S. 118 ff) sind Öffentlichkeit und Transparenz hier ein Muss und Programm von Mercedes-Benz, entsprechend dynamisch gestaltet sich der Raumfluss. Im Prinzip bildet die 22 m hohe Halle einen einzigen Raum. Auf Galerien, die gleichzeitig auch die Auf- und Abfahrtsrampen sind, stehen die Objekte der Begierde, häufig mit Preisen im sechsstelligen Eurobereich. Die Gesamtnutzfläche beträgt 35.000 qm. Grundriss und Rampen sorgen für Bewegung, die Architekten sprechen gern von einer „Landschaft" unter Dach, die mit einem 750 Jahre alten Olivenbaum, einer alpintauglichen Kletterwand und einer Bühne über Wasser mit Wasserwänden bestückt ist. Das Gebäude wirkt technoid; durch die Vielzahl der internen Orte und Plätze auf unterschiedlichen Niveaus, durch Grün im Haus und das fließende Licht ergibt sich eine großstädtische Sinfonie der Eindrücke und Signale, verdichtet auf ein einziges Bauwerk.

Baujahr	2000
Nutzfläche	35.500 qm, Ausstellungsfläche: 14.000 qm
Sonderattraktionen	Bühne, Kletterwände, Themengärten, Restaurant, Formel-1-Fahrsimulator, Bobbycartbahn, Oldiekino, LED-Videowand, Oval Office mit Bar
Architekten	Lamm-Weber-Donath und Partner, Stuttgart

In complete contrast to the Sindelfingen Design Center (p. 118 ff), publicity and transparency are a must here, part of Mercedes-Benz policy, and the flow of the building is correspondingly dynamic. Basically, the 22 metre-high hall is one single room. The objects of desire, frequently with price tags running to six-figure euro sums, are arranged on galleries, which are also the up and down-ramps. Total utilizable floor space comes to 35,000 m². The ground plan and the ramps create an impression of movement, and the architects like to speak of a roofed-over "landscape", complete with 750-year-old olive tree, Alpine climbing wall and a stage over water with waterfall walls. The building is technoid in character, and the many internal places and squares on different levels, together with the greenery in the building and the flow of light result in a metropolitan symphony of impressions and signals, condensed in one single building.

Constructed	2000
Usable space	35,500 m², exhibition space: 14,000 m²
Special attractions	Stage, climbing walls, theme gardens, restaurant, Formula One driving simulator, go-cart track, oldies cinema, LED video wall, oval office with bar
Architects	Lamm-Weber-Donath und Partner, Stuttgart

Der Gebäudecharakter ist technoid, durch das fließende Licht ergibt sich eine großstädtische Sinfonie der Eindrücke.
The building is technoid in character, and the greenery in the building and the floating light combine to form a metropolitan symphony of impressions.

gebaute kommunikation: bmw und mini auf der iaa 2001

Wenn vierrädrige Ikonen auf einer Autoausstellung still stehen müssen, dann soll wenigstens die architektonische Hülle dynamisch sein. Das war die Vorgabe von BMW zur IAA 2001 an den Frankfurter Architekten Bernhard Franken (in Arbeitsgemeinschaft mit ABB Architekten Frankfurt). Der entwickelte eine entsprechende „Dynaform", und die wirkt außen dank des gekurvten und gefalteten, faserverstärkten PVC wie ein riesiger weißer Lindwurm. Auch innen ist hier nichts rechtwinklig und gerade, alles fließt, und so wird der 130 m lange Pavillon zu einer überdachten Straße im Tunnel, und die Autos darin markieren das Fixierbild eines virtuellen Autorennens.

Der Showroom ist der Held einer Bewegung, mit einer Architektonik, die aus dem Datensatz kommt und deren komplexe Geometrien Euklid und Archimedes wie Erstklässler aussehen lassen. Aus der „Mastergeometrie" entstehen logische Ableitungen – Derivate –, sie definieren die Neigungen der Böden, die Formen der Stahlträger, deren Ausrichtung, die Krümmungen des Baukörpers. Alle Details, wie die Standpunkte von Show-Electronic, Screen, Boxen und Autos, wurden per Computer in einem x-/y-/z-Koordinatensystem rechnerisch festgelegt.

Aber hoher Aufwand allein schützt nicht vor Kritik. Die sehr lässige Bauausführung brachte der Dynaform das Prädikat „Bierzelt" ein (*Die Zeit*). Diesen Eindruck zu ändern, könnte auf der nächsten IAA gelingen, wenn der inzwischen demontierte und eingelagerte „Lindwurm" wiederauferstehen wird.

Dass man die gesicherte architektonische Welt des rechten Winkels nicht vergessen hat, beweisen die Architekten mit dem Juniorpavillon für den neuen Mini: der ist klassisch, ja streng, kubisch – und das für ein Auto, das der jungen und jung gebliebenen Generation gehört.

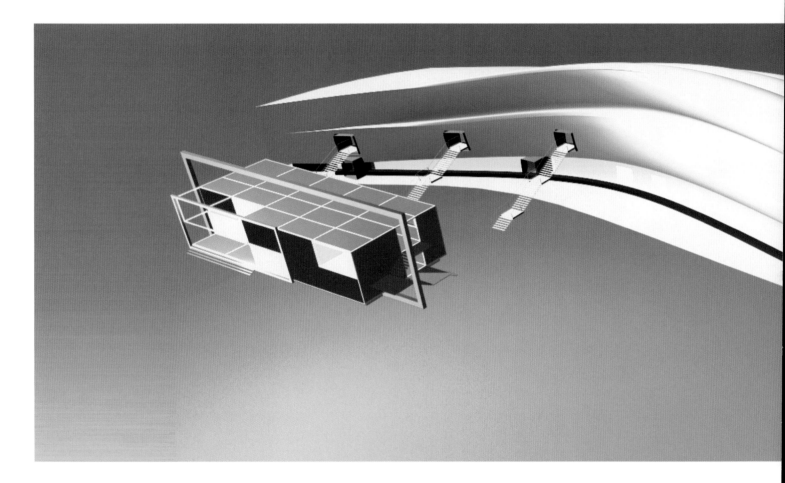

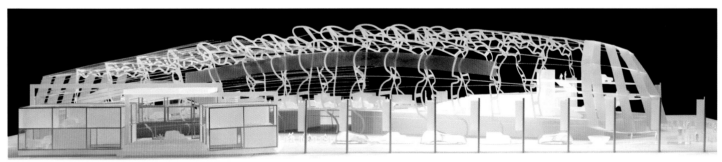

Kommt dem Original sehr nahe: Arbeitsmodell für die Dynaform.
Very close to the original: the working model for the dynaform.

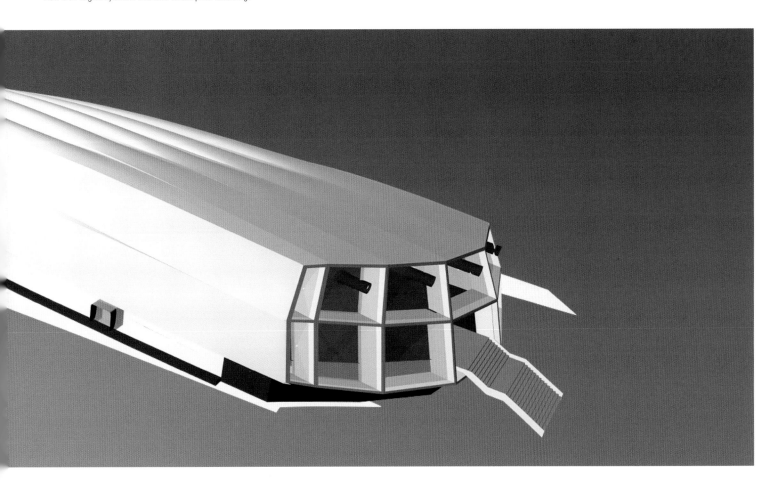

structured communication: bmw and mini at the iaa 2001

If four-wheeled icons have to stand still at a car show, then at least the architectural wrapping should be dynamic. That was the brief the Frankfurt architect Bernhard Franken (in collaboration with ABB Architekten Frankfurt) received from BMW for the IAA 2001. He developed a corresponding "dynaform" whose outward appearance, thanks to a curved and folded fibre-reinforced PVC sheet, is that of a huge white lindworm. On the inside, too, there are no rectangles or straight lines, everything is in movement, and so the 130 metre-long pavilion is like a roofed-over road in a tunnel, and the cars in it like a snapshot of a virtual car race.

The showroom is the hero of a movement whose architectural technique is computer-generated, and whose complex geometries make Euclid and Archimedes look like absolute beginners. Logical derivatives arise out of this "master geometry", defining the gradient of the floors, the forms of the steel beams and their alignment, and the curves of the building unit. All the details, such as the location of the entertainment electronics, the screen, the pits and the cars, were calculated by computer in a system of x, y and z coordinates.

But all this hard work is no protection from criticism. The extremely careless construction work led critics to describe the dynaform as a "beer tent" (*Zeit*). It may be possible to change this impression at the next IAA, when the "lindworm" that has meanwhile been disassembled and stored away is resurrected.

The architects prove that they have not forgotten the received architectural world of the right angle in their junior pavilion for the new Mini: it is classical, if not severe and cubist – unexpected qualities for a car that belongs to the young and young-at-heart generation.

Aufbau der Dynaform: die PVC-Folie wird über die Konstruktion gespannt.
Erecting the dynaform: the PVC sheet is stretched over its frame.

Impressionen von der IAA Frankfurt 2001 (s. auch folgende Seiten)
Impressions of the IAA Frankfurt 2001 (see also following pages)

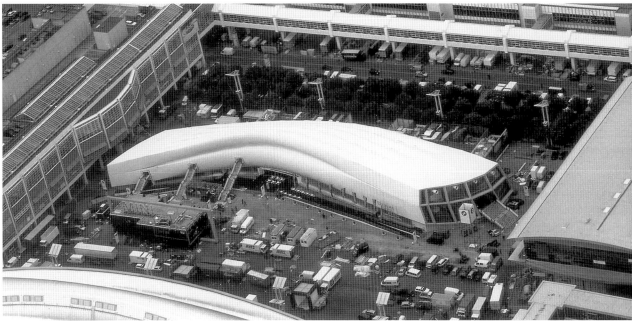

Willkommen.
Welcome.

Dynaform innen, der Dynamik einer Straße nachempfunden.
Inside the dynaform, based on the dynamism of the open road.

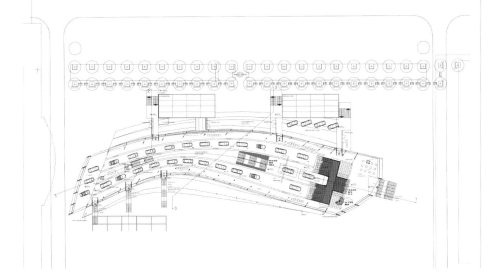

Baujahr	**2001**
Architekt	**Bernhard Franken, Frankfurt**
Constructed	2001
Architect	Bernhard Franken, Frankfurt

Grundriss und Lageplan auf der IAA: Dynaform und Cube liegen neben der neuen Messehalle von Nicholas Grimshaw.
Ground plan and layout at the IAA: dynaform and cube next to Nicholas Grimshaw's new trade fair building.

Kontrapunkt: der Juniorpavillon für den neuen Mini – klassisch, streng, kubisch.
Counterpoint: the junior pavilion for the new Mini – classical, severe, cubist.

Den BMW-Pavillon auf der IAA 2001, „Dynaform", halten Sie nicht für Architektur, sondern für „gebaute Kommunikation".

Er markiert die Grenzlinie zwischen Architektur und Kommunikation. Wir bauen für die Marketing-Abteilung, für die die „Message" wichtiger ist als das Bauwerk – also gehen wir anders vor.

Und das nicht zum ersten Mal. Sie haben schon auf der IAA 1999 für BMW einen Pavillon gebaut und im Rahmen der EXPO 2000 einen „Blob" in München entworfen. Alle diese Beispiele definieren sich durch eine raffinierte Leichtigkeit – nicht Statik charakterisiert sie, sondern Dynamik. Häuser wie Anzeigenkampagnen?

Natürlich ist „Dynaform" am Schluss ein Gebäude als Schutz gegen Wind und Wetter. Doch am Anfang steht der Auftrag, ein neues Auto zu präsentieren. Das Fatale an Messen ist ja, dass Autos hier still stehen müssen. Deswegen wollen wir den Raum drumherum so beschleunigen, dass ein Gefühl des Fahrens entsteht – Dynaform eben.

So ein Projekt wird nicht mehr auf dem Papier entworfen, sondern mit einem hochintelligenten Computerprogramm. Mit welcher intuitiven Vorgabe betraut der Architekt den Rechner?

Wir haben uns am Doppler-Effekt orientiert, das ist der Klang eines Rennautos, der, wenn es auf mich zukommt, gedrückt, und wenn es mich verlässt, wieder gedehnt wird. Also erst höher, dann tiefer zu hören ist. Diesen Effekt übertragen wir auf den Raum. Dazu benutzen wir eine Film-

animations-Software, mit der Filmleute in Hollywood z.B. die Titanic untergehen lassen. Wir benutzen einfach die physikalischen Gesetze, die in diesem Programm implementiert werden, lassen sie auf unseren Entwurf einwirken und erzählen damit eine bestimmte Geschichte.

Wie macht man aus diesem Computer-Experiment ein richtiges Haus?

Wir verfolgen das Konzept von digitaler Durchgängigkeit: Wir entwerfen am Computer und versuchen, mit rechnergestützten Methoden das Design durchzusetzen. Am Anfang steht die „heilige Mastergeometrie", ein Datensatz, der eine genaue Geometrie in 3D festlegt. Dieser Datensatz mit allen seinen Derivaten, also Ableitungen, steht dann im Internet und wird von allen beteiligten Planern und Ingenieuren bearbeitet. Das reicht dann bis zu den Programmteilen für die Werkzeuge, mit denen aus den entsprechenden Werkstoffen die verlangten Flächen und Formen auszuschneiden sind. Bei diesem Projekt reden wir von bis zu 30.000 verschiedenen Geometrien.

Woraus besteht nun der Pavillon?

Aus Stahl und einer Membranhaut.

Lohnt sich der gigantische Aufwand dafür, dass man merkt, dass es eine Welt jenseits des rechten Winkels gibt?

Ja, wir haben herausgefunden, dass man Dynamik und Kräfte körperlich im Pavillon spüren kann. Das ist genau das, was wir erreichen wollen.

→ **interview** interview

Bernhard Franken, junger Frankfurter Architekt, ist Vefechter der computergenerierten Architektur.
Bernhard Franken, a young Frankfurt-based architect, is a champion of computer-generated architecture.

In your view, "dynaform", the BMW pavilion at the IAA 2001, is not architecture but structured "communication".

It defines the borderline between architecture and communication. We are building for the marketing department that wants to convey a message and is not so interested in a piece of architecture – and so our approach is a different one.

Not for the first time, either. You built a pavilion for BMW at the IAA in 1999, and designed a "blob" in Munich in connection with EXPO 2000. All these examples are characterized by sophisticated lightness – their typical feature is dynamism, not stasis. Buildings like advertising campaigns?

Of course, when all is said and done, "dynaform" is a building offering protection from wind and weather. But the starting point was our brief to present a new car. What is so aggravating about trade fairs is that cars have to stand still there. That is why we want to accelerate the space around them, so that a driving feeling is created – that's what we mean by dynaform.

A project like this is no longer drafted on paper, but with the help of highly intelligent software. What intuitive task does the architect entrust to the computer?

We took our lead from the Doppler effect, from the sound a racing car makes: compressed as it comes towards me and stretched again as it drives away from me. In other words, its pitch is first higher then lower. We transfer this effect

to space. To do this, we use the sort of film animation software that Hollywood film people use to make the Titanic sink. We simply employ the physical laws that are implemented in this program, allow them to work on our design, and in this way tell a certain story.

How do you make a real building out of this computer experiment?

The concept we pursue is one of digital consistency: we design at the computer and try to apply the design using computer-assisted methods. The starting point is the "holy master geometry", a data record which fixes an exact geometry in 3-D. This data record, together with all its derivatives, is then posted on the internet and processed by all the planners and engineers involved. This encompasses the sub-programs for the tools to be used to cut the surfaces and forms required from the corresponding materials. In a project like this, we're speaking about as many as 30,000 different geometries.

What is the pavilion made of?

Steel and a membrane skin.

Is it worth all this colossal effort just to see that there is a world beyond the right angle?

Yes, what we have found is that you can physically feel the dynamism and forces of the pavilion. And that is precisely what we want to achieve.

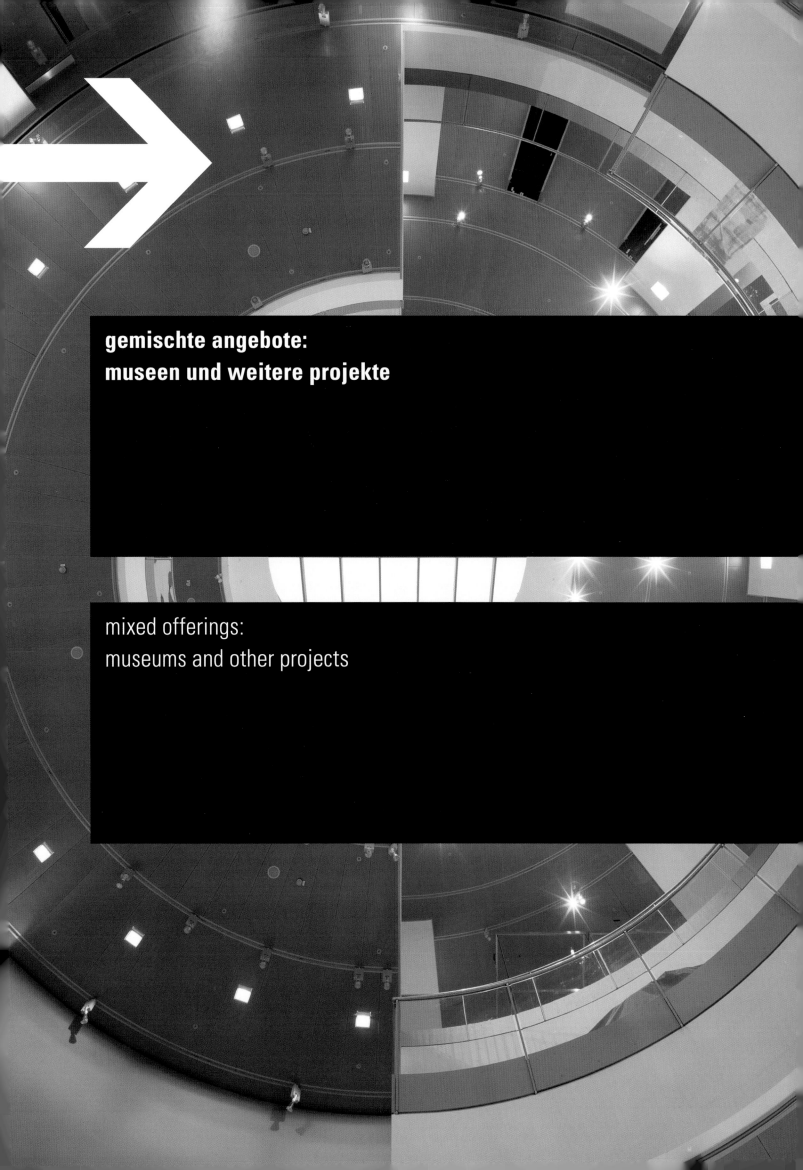

**gemischte angebote:
museen und weitere projekte**

mixed offerings:
museums and other projects

leinwand für voyeure und flaneure: smart city, hamburg

Der Friedrich-Ebert-Damm in Hamburg-Wandsbek hat sich in den achtziger Jahren zu einem typischen Autostrip amerikanischer Prägung
entwickelt. Autohäuser und Werkstätten aller wichtigen Marken reihen
sich brav und austauschbar an diesem Vorstadt-Boulevard auf, dem es an
einem eigenen urbanen Ausdruck fehlte.

Spätestens mit dem Stützpunkt für smart, der Automarke der ökobewuss-
ten Yuppies, hat sich das geändert, denn der frivole neue Kleinwagen
begnügte sich nicht mit einem Showroom, sondern forderte zusätzlich
Einrichtungen für Entertainment und Infrastruktur tatsächlicher und mög-
licher Kunden ein, die alle zusammen in einem Gebäudekomplex organisiert wurden. Das hat zum längsten Hamburger „Lichtspielhaus" geführt,
denn hier liegen nun hinter gläserner Front – abends mit Beleuchtung –
Multiplex-Kino, Restaurants, Fitness-Studio und vieles mehr. Diesem
urbanen Nutzungsfächer gestanden die Architekten kräftige Baukörper
zu: Riegel und Turm, einfache, aber stadtraumbildende Elemente, die bei
der schnellen Vorbeifahrt im Auto wahrnehmbar bleiben und die übrige
„Schrottplatz-Optik" eines Autostrips geschickt verdecken und über-
spielen. Der Turm dient der Performance des smart und ist inzwischen
zum Hingucker dieser „Auto-Straße" geworden. >

Bauzeit	1998–1999
Maße	Länge: 168 m; Höhe: 18 m; smartturm: 24 m
Konstruktion	Stahlbetonskelett mit punktgehaltener Glasfassade
Architekten	Von Gerkan, Marg und Partner, Hamburg und
	Schildt Architekten, Hamburg

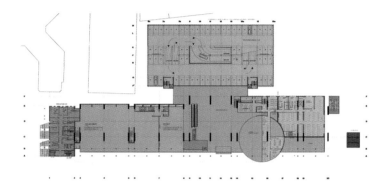

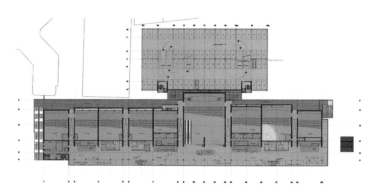

Grundrisse Plans

screen for voyeurs and flaneurs: smart city, hamburg

In the 1980s, Friedrich-Ebert-Damm in Hamburg's Wandsbek district developed
into a typical US-style strip, with car salesrooms and workshops for all the
major marques lining up in uniform style along a suburban boulevard that was
devoid of any urban expression.

All this has changed since the base for the smart arrived, the marque for ecolo-
gically-conscious yuppies: this frivolous new small car was not satisfied with
a showroom, but demanded extra facilities for entertainment and infrastructure
for actual and potential customers, all of which was organized together in one
complex. The result is Hamburg's longest "moving picture show", for behind its
glass façade, which is lit up at night, you will find a multiplex cinema, restaurants, a gym and lots more. The architects have set this range of urban uses in
powerful building elements. Oblong and tower: simple elements, yet ones
that create urban space, which remain perceptible when driving past at speed
and cleverly conceal and play down the usual "scrap yard feel" of a car strip.
The tower is part of the performance staged for the smart, and has meanwhile
become the focal point of this "automotive street". >

Constructed	1998–1999
Dimensions	Length: 168 m; Height: 18 m; smart tower: 24 m
Structure	Reinforced steel skeleton with glass façade, supported pointwise
Architects	Von Gerkan, Marg und Partner, Hamburg and
	Schildt Architekten, Hamburg

**Stadt unter Dach: Der Showroom von smart und die Werkstatt sind in eine Landschaft
aus Gastronomie, Entertainment und Wellness eingebettet.**
Roofed-over town: The smart showroom and its workshop are embedded in a landscape of
gastronomy, entertainment and wellness.

**Länge läuft: die 168 m lange Fassade der smart city veränderte das
Gesicht eines Hamburger Autostrips nachhaltig.**
Length sells well: the 168 metres of smart city's façade have permanently
changed the face of one of Hamburg's strips.

Der Riegel nimmt vor allem die Lobby mit den einzelnen Sälen des Multiplex-Kinos auf und wurde mit einer der größten Planar-Fassaden Deutschlands voll verglast. Er wird zu einer gigantischen Stadt-Leinwand auf 168 x 18 m für Voyeure und Flaneure draußen und drinnen; für die „Schauspieler", die in Wirklichkeit Kinobesucher in den Foyers und Bars sind oder Menschen, die gerade ein Auto kaufen. Kurzum, es findet ein Live-Spektakel statt, das signa-

lisiert: Hier ist niemand allein, hier liegt ein komprimiertes Stück Stadt. Die Architekten konfrontieren die präzise, technische Fassade mit einer dahinter liegenden Kulissenarchitektur, die in ihrer Buntheit die Welt der englischen 60er-Jahre karikiert. Es startet das Raumschiff „Friedrich-Ebert-Damm" zusammen mit dem smart, der Ikone der juvenilen Mobilität.

Lobbies und Restaurants liegen hinter einer der
größten Planar-Fassaden Deutschlands.
Lobbies and restaurants behind one of the largest
Planar façades in Germany.

Above all, the oblong houses the reception area together with the individual cinemas of the multiplex complex, and its fully glazed front is one of Germany's largest Planar façades. Its role has become one of a giant urban screen measuring 168 by 18 metres, for the benefit of voyeurs and flaneurs inside and out; for the "actors" who are in reality cinema-goers in the foyers and bars, or people in the act of buying a car. In short, it is a live event signalling that nobody is alone here, for here is a condensed slice of urban life.

The architects create a contrast between the precise, technological façade and the stage scenery lying behind it, whose colourfulness is a caricature of the swinging English scene in the 1960s: welcome to the launch of the starship "Friedrich-Ebert-Damm" together with the smart, the icon of youthful mobility.

salon mit gartenzimmer: audi forum, ingolstadt

„Das Rad ist rund", schrieb Peter Rumpf in der *Bauwelt* und hat damit das „museum mobile", das Automuseum der Audi-Stadt Ingolstadt, trefflich beschrieben. Der Kreisgrundriss des Museums und die transparente Glasfassade sind die Fixpunkte des Ausstellungskonzepts in Ingolstadt, die Helden der Audi-Mobilität sind also auch von der Straße her sichtbar.

Das gesamte Ensemble des Audi Forum ist ein wunderbares Glacis für 32.000 Arbeitsplätze. Denn ein Werkstor ist im Medienzeitalter als auschließlicher Grenzposten für den Übergang zwischen Stadt und Werk nicht mehr tauglich. Es existiert natürlich noch, aber davor ist jetzt eine Marketingzone mit Piazza und Piazzetta, mit Museum und Kundenzentrum entstanden. Letzteres adelt die Funktion der Fahrzeugübergabe auf interessante Weise, indem es mit einer Bautypologie eines anderen schnellen Verkehrsmittels kokettiert: dem Flugzeughangar. Insgesamt ein „Salon mit Gartenzimmer" (Benedikt Loderer) statt des früheren Nadelöhrs, das Werkstor heißt. Automobilhersteller lernen schnell.

Bauzeit	1999–2000
Konstruktion	Stahlbeton mit selbsttragender Pfosten-Riegel-Glasfassade
Bruttogeschossfläche	Museum: 6.500 qm, Kundencenter: 8.000 qm, Verbindungsbau: 3.500 qm
Architektur und Gesamtkonzept	Henn Architekten, München
Ausstellungs- und Kommunikationsdesign	KMS Team, München
Freiraumplanung	Vittorio Magnago Lampugnani, Mailand mit Marlene Dörrie und Jens Bohm

salon with a conservatory: audi forum, ingolstadt

"The wheel is round", Peter Rumpf wrote in the magazine *Bauwelt*, providing a fitting description of the "Museum Mobile", as the car museum in Ingolstadt (where Audi has its HQ) is called. The circular plan of the museum and the transparent glass façade are the points of reference of the Ingolstadt exhibition concept, and so the heroes of Audi-mobility are also visible from the road outside.

Audi Forum's overall ensemble is a wonderful glacis for 32,000 workplaces. After all, in the mass media age a factory gate is no longer appropriate as the sole frontier post guarding the transition between town and plant. Of course, it still exists, but in the area in front of it we now have a marketing zone with a piazza and piazzetta, with a museum and a customer centre. The latter upgrades the function of vehicle hand-over in an interesting way by flirting with the architectural typology of another means of rapid transport: the aircraft hangar. The whole complex, to quote Benedikt Loderer, is a "salon with a conservatory", instead of the former bottleneck known as a factory gate. Car manufacturers are fast learners.

Constructed	1999–2000
Structure	Reinforced concrete with self-supporting column and beam glass façade
Gross floor area	Museum: 6,500 m^2, customer centre: 8,000 m^2, connecting building: 3,500 m^2
Architecture and overall concept	Henn Architekten, Munich
Exhibition and communication design for the museum	KMS Team, Munich
Planning of open spaces	Magnago Lampugnani, Milan with Marlene Dörrie and Jens Bohm

Rund und transparent: Kernstück des Ingolstädter Audi Forums an der Nahtstelle zwischen Stadt und Werk ist das Automuseum.

Round and transparent: The core of Ingolstadt's Audi Forum, at the interstice between town and plant, is the car museum.

In der Kontur eines Flugzeughangars: Ort der Fahrzeugübergabe in Ingolstadt.
The contours of an aircraft hangar: where cars are handed over in Ingolstadt.

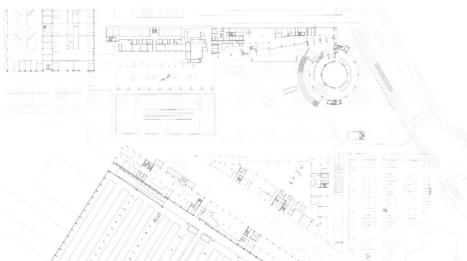

neuer stern über stuttgart:
mercedes-benz museum, untertürkheim

Mercedes-Benz als DaimlerChrysler-Marke ist in Stuttgart omnipräsent. Eine alles überstrahlende Figur, sozusagen das Bauwerk, mit dem der Konzern auf den Punkt kommt, fehlt allerdings noch: Das Neue Mercedes-Benz Museum, das seit 2003 in Untertürkheim entsteht, wird dieses Defizit beheben. In einem internationalen Wettbewerb hat sich das niederländische UN Studio des Architekten Ben van Berkel durchgesetzt und einen skulpturalen Bau vorgesehen: eine 48 m hohe Doppelhelix. Es ist ein Museum neuer Typologie, nicht für die Kunst, sondern für die Autokunst, den Mythos Mercedes. Man kann es eine raffinierte Neuauflage des legendären Guggenheim-Museums in New York von Frank Lloyd Wright nennen. Die Ausstellungsflächen mäandern von unten nach oben, Ben van Berkel bezeichnet es als „eine Landschaft der Mobilität". Dieses niederländische Büro ist seit seinem Möbius-Haus in Holland Spezialist für eine neue Raumauffassung, bei der die fließenden Teile so überraschend komponiert werden, dass kaum noch zwischen innen und außen, zwischen Dach und Wand unterschieden werden kann. Die richtige Rundum-Architektur für den Mythos und die Geschichte eines Autoherstellers, dem vom Silberpfeil-Rennwagen bis zum Mega-Truck alles gelungen ist.

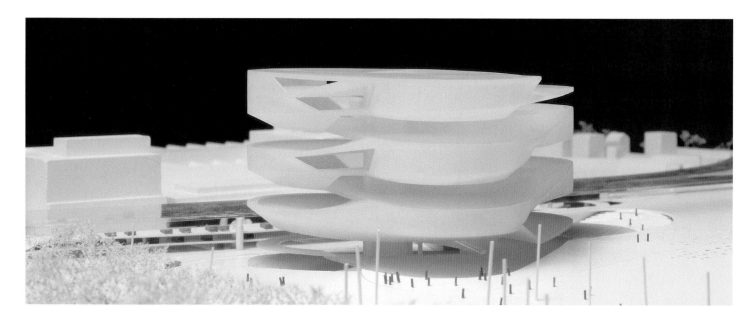

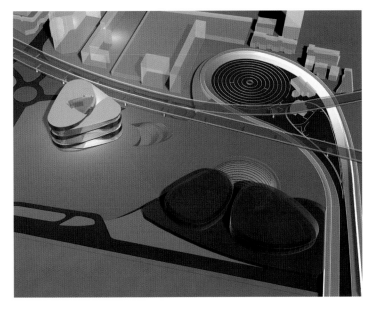

new star over stuttgart:
mercedes-benz museum, untertürkheim

DaimlerChrysler's Mercedes-Benz marque is omnipresent in Stuttgart. But there is still no one feature outshining all the rest; a building with which the company makes its point, as it were. This shortcoming is to be remedied by the New Mercedes-Benz museum, the construction of which began in 2003. Ben van Berkel's Dutch UN Studio was given the contract following an international competition. Their design calls for a sculptural construction: a 48 metre double helix. It is a new type of museum, not for art but for automotive art, for the Mercedes myth. One could say it was a sophisticated new version of Frank Lloyd Wright's legendary Guggenheim Museum in New York. The exhibition surfaces meander upwards in what Ben van Berkel calls a "landscape of mobility". Since their Moebius House in Holland, this Dutch team have specialized in a new concept of space in which the flowing elements are composed so surprisingly that it is scarcely possible any longer to distinguish between inside and out, between roof and wall: just the right "all-round" architecture for the myth and history of a car manufacturer with the Midas touch, from the Silver Arrow racing car to the giant truck.

Neuer „Stern" für die Autostadt Stuttgart: ein silberglänzendes Automuseum als „Landschaft der Mobilität" (Ben van Berkel).
New "star" for the motown Stuttgart: a gleaming silver car museum designed as a "landscape of mobility" (Ben van Berkel).

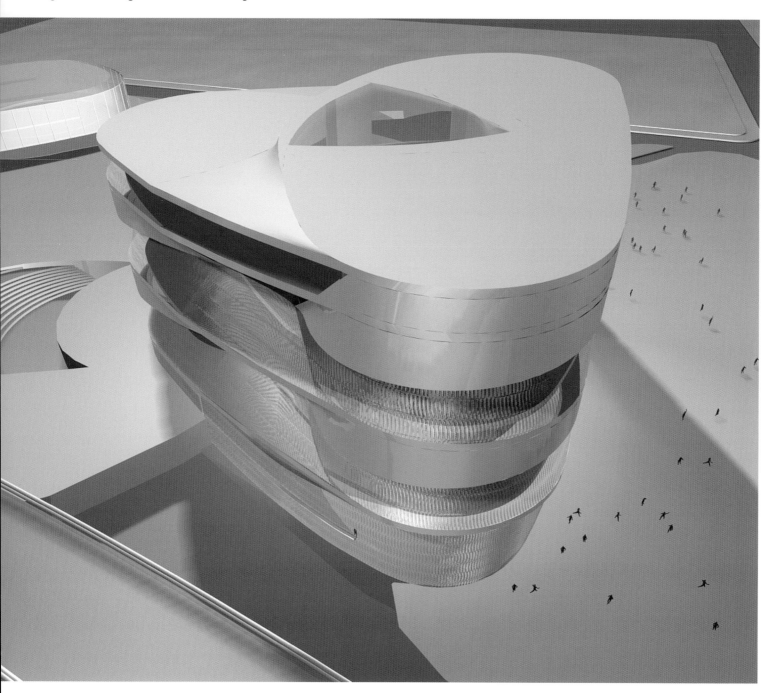

Rundum-Architektur für den Mythos Mercedes-Benz in einer Architekturqualität, die an das Guggenheim-Museum von Frank Lloyd Wright in New York erinnert.
For the Mercedes-Benz myth, curved, ring-like architecture of a quality reminiscent of Frank Lloyd Wright's Guggenheim Museum in New York.

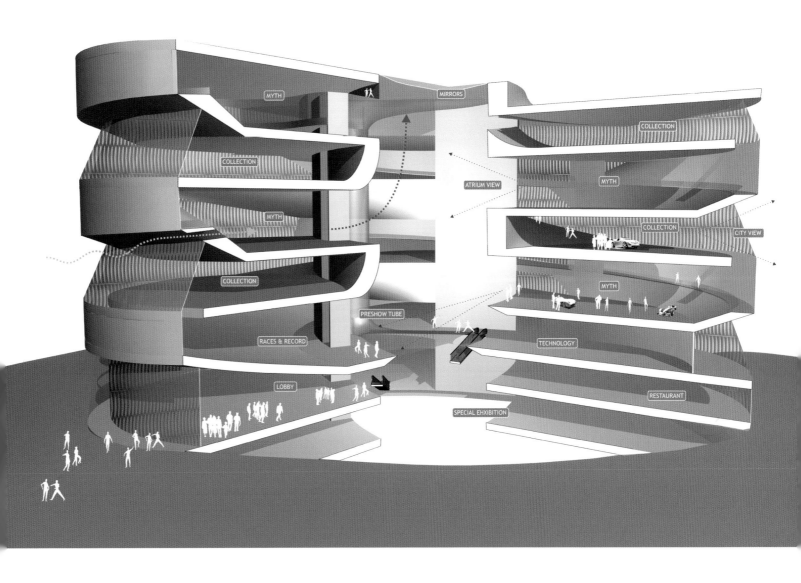

Labels in image 1:
MYTH · MIRRORS · COLLECTION · COLLECTION · MYTH · ATRIUM VIEW · MYTH · COLLECTION · CITY VIEW · COLLECTION · MYTH · PRESHOW TUBE · RACES & RECORD · TECHNOLOGY · LOBBY · RESTAURANT · SPECIAL EHXIBITION

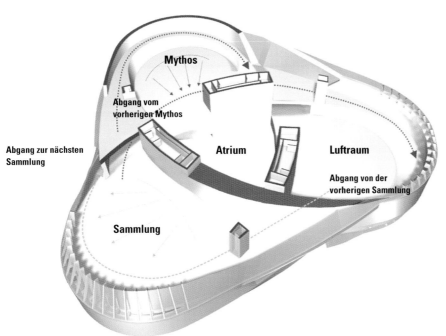

Labels in image 2:
Mythos · Abgang vom vorherigen Mythos · Abgang zur nächsten Sammlung · Atrium · Luftraum · Abgang von der vorherigen Sammlung · Sammlung

**Grundrissschema und Schnitt
durch das neue Ausstellungsgebäude.**
Ground plan and section
of the new exhibition building.

high tech aus tradition: bugatti-manufaktur, molsheim

Muster mit Wert: Die elegante Architektur des Château Saint-Jean aus dem Jahr 1857 (diente früher dem Malteserorden) steht unter Ensembleschutz und wird zum Gastgeber für Bugatti – es war einmal das Gästehaus dieses Autopioniers. Aufgrund des lädierten Zustands im Innern wurde das Gebäude entkernt, tiefer gelegt und mit einer „weißen Wanne" ausgebildet, modern, bautechnisch Spitze, aber mit einem traditionellen Gesicht. So ähnlich ist das Projekt Bugatti im VW-Konzern zu verstehen: Tradition erkennen, bewahren und auf den Weg in die Zukunft bringen. Repräsentation, Verkauf, Ausstellung und Administration finden im Haupthaus statt. In den ehemaligen Remisen und weiteren (auch neu gebauten) Nebengebäuden werden alte Bugatti restauriert und neue an die Kunden übergeben. Nebenan wird auf einem 80.000 qm großen Gelände ein neues Bugatti-Montagewerk entstehen.

high tech by tradition: bugatti factory, molsheim

A valuable pattern: the elegant architecture of the Château St. Jean, built in 1857 (it formerly housed the Order of the Knights of St. John) and a classified historic building, has become the home for Bugatti – it used to be the guesthouse of this car pioneer. Because the interior was in a dilapidated state, the building was gutted, its floor lowered and lined with a "white tub". This was modern and architecturally progressive, yet preserved the building's traditional appearance. This is similar to the way the Bugatti project has to be seen within the VW group: acknowledging and preserving tradition, and setting it on course for the future. Reception areas, sales, exhibitions and administration are located in the main building. In the former outbuildings and other annexes (some of them newly built), old Bugattis are restored and new ones handed over to their buyers. Next door, on a plot measuring 80,000 m², a new Bugatti assembly plant is to be built.

Klassisch und modern: Das Schloss passt exakt zur programmatischen Beschreibung einer neuen Generation von Bugatti-Fahrzeugen.
Classical and modern: Molsheim's château could not better fit the programmatic description of the new generation of Bugatti vehicles.

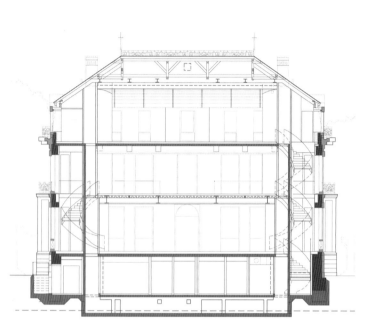

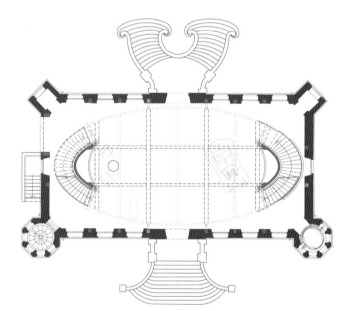

Links: Schnitt und Grundriss EG des Château Saint-Jean: Raum für Repräsentation und Show. Oben und rechts: neue Nebengebäude als Heimat für die Restaurierung alter Bugatti und die Übergabe neuer Fahrzeuge.

Left: Section and plan of the ground floor of Château St. Jean: room for receptions and shows. Top and right: new annexes where old Bugattis are restored and new ones handed over to their buyers.

Bauzeit	**ab 2002**
Architekten	**Henn Architekten, München**

Start of construction	2002
Architects	Henn Architekten, Munich

kraftfeld mit strahlenbündeln: bmw-werk, leipzig

Die Ideen aus Dresden und Mladá Boleslav haben neue Impulse für den Bau von Automobilwerken gegeben. Zaha Hadid, die einen Wettbewerb für den zentralen Bereich des neuen BMW-Werks in Leipzig gewonnen hat, setzt diese Erkenntnisse allerdings auf eigene, individuelle Weise um: „Mich interessieren Verkehrsströme, denn Bewegung verändert die Wahrnehmung", sagt sie.

Wenn Zaha Hadid Transparenz schafft und Grenzen zwischen Fertigung und Verwaltung aufhebt, tut sie das auf schnittigen Grundrissen, auf Kraftfeldern, die aus Überlagerungen verschiedener Funktionen entstehen. Hadid nennt das Zentralgebäude für Auslieferung, Erlebnisbereiche, Kommunikation und Verwaltung „Kompressionszone". Die eigentliche Fertigung mit ihren Fertigungsstraßen oder -strahlenbündeln wächst aus diesem „Nervenzentrum" heraus, als müsste sie bildlich das zentrale Kraftfeld widerspiegeln.

Zaha Hadids Leistung besteht darin, dass sie die Dynamik dieser Bauaufgabe aus Industrie, Gewerbe und Kultur nicht nur symbolisch umsetzt, sondern auch die funktionalen Zusammenhänge optimiert. Schließlich ist das Werk bei dieser Konzeption einfach zu erweitern oder umzustrukturieren, was bei der weiterhin dynamischen Entwicklung der Autoindustrie eine Conditio sine qua non ist.

force field with focussed beams: bmw factory, leipzig

The ideas formulated in Dresden and Mladá Boleslav have provided new impetus for the building of new car plants. But when Zaha Hadid, who was awarded the contract for the central area of the new BMW factory in Leipzig, puts these insights into practice, she does so in a uniquely individual way: "I'm interested in flows of traffic, because movement alters perception", she says.

When Zaha Hadid creates transparency and abolishes the borderlines between shop floor and management, she makes use of snappy ground plans and of force fields that arise out of the overlapping of different functions. Hadid calls her central building for dispatch, event zones, communication and administration a "compression zone". Production itself, with its production lines (or rather focussed beams), grows from this "central nerve", as though it had to figuratively reflect the central force field.

Zaha Hadid's achievement is that she has not only provided a symbolic representation of the dynamism of the building task, with its industrial, commercial and cultural features, but has also optimized their functional interrelationships. Finally, her design allows the factory to be extended or reorganized simply – a conditio sine qua non, given that the development of the car industry will remain a dynamic one.

Planung für das neue BMW-Werk Leipzig, Zentralgebäude von Zaha Hadid.
Design of the new BMW factory in Leipzig, with Zaha Hadid's central building.

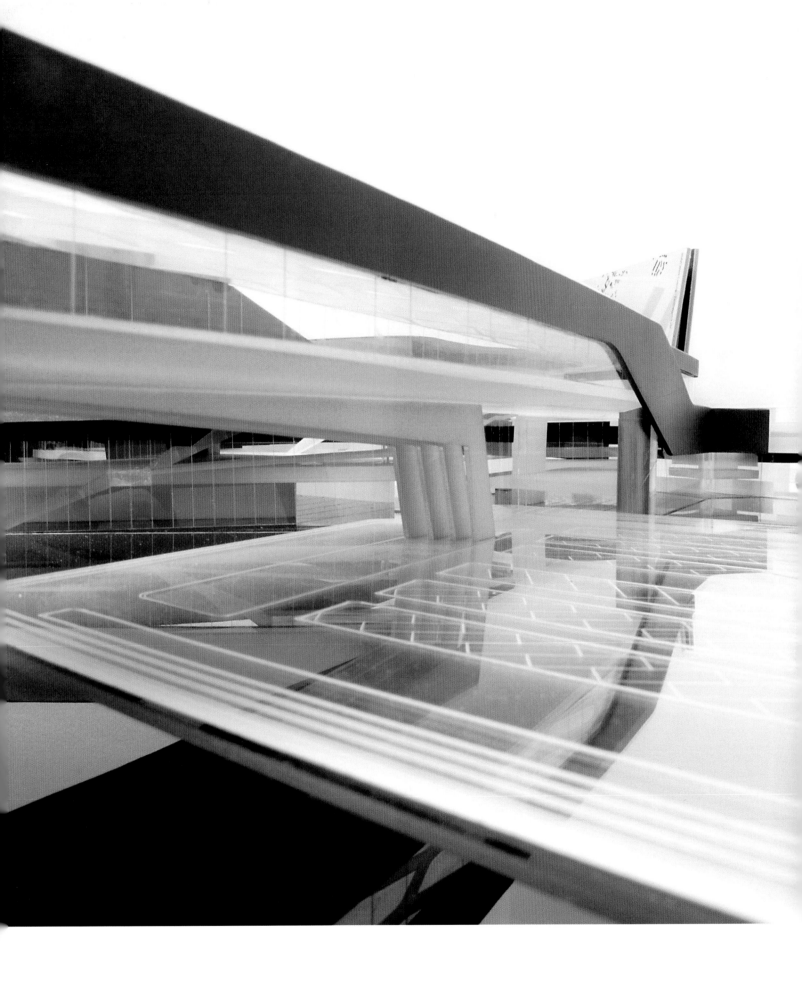

Innenansichten des Zentralgebäudes und Modell
Interior view of the main building and model

clean energy cloud: bmw welt, münchen

Die BMW Welt ist „unser Portal zum Dialog mit der Öffentlichkeit", sagte der frühere BMW-Vorstandsvorsitzende Prof. Joachim Milberg. 27 Starteams aus Los Angeles und New York und ganz Europa wurden zu einem Architektenwettbewerb eingeladen. Wenn schließlich der Entwurf der Wiener Gruppe Coop Himmelb(l)au sich als Erster durchgesetzt hat, dann weil sein Dach, ja seine Dachlandschaft wie eine Himmelslandschaft grenzenlose Freiheit verspricht. Die Jury schrieb: „Die Arbeit lebt von einer großen Idee: ein Marktplatz unter einem weiten Dach. Dieses Bild assoziiert Offenheit und Kommunikation". >

clean energy cloud: bmw welt, munich

Former chairman of the BMW board Prof. Joachim Milberg commented that the BMW Welt (BMW World) was "our portal for dialogue with the general public". Twenty-seven teams of star architects, from Los Angeles, New York and all over Europe, were invited to compete for the project. The reason why the contract was finally awarded to the Vienna-based Coop Himmelb(l)au was that their roof, or rather roof landscape, is like a skyscape holding out the promise of unrestricted freedom. As the judges wrote: "This work is informed by a grand design: a marketplace under a wide roof. This image has associations of openness and communication". >

Geniale Architekturantwort auf das BMW-Hochhaus (Architekt Karl Schwanzer) und das Münchner Olympiagelände (Günter Behnisch, Frei Otto): die BMW Welt.
Congenial architectural response to the BMW skyscraper (by Karl Schwanzer) and Munich's Olympic grounds (Günter Behnisch, Frei Otto): BMW Welt.

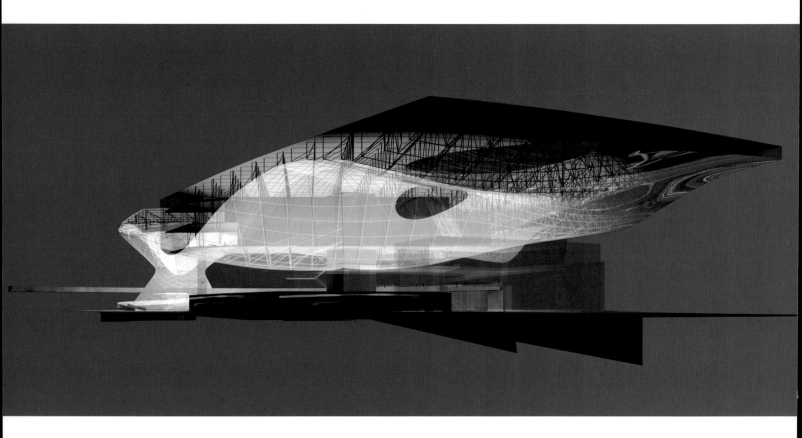

Dieses Haus wird ein Dach sein – ohne „eigentliche Fassaden, also eine skulpturelle Einheit", so Coop Himmelb(l)au, und man wird das Uraltthema von Balken und Stütze ablösen und Neuland betreten.

Der Frontman von Coop Himmelb(l)au, Wolf Dieter Prix, hat dazu eine blumig-griffige Bezeichnung zur Hand: „Clean Energy Cloud" nannte er das intelligente Dachsystem, das nach der Überarbeitung zweischalig Belüftung, Kühlung und Energiegewinnung regeln wird. Selbst ein Solarkraftwerk könnte integriert werden. Was von der phantastischen Kombination aus Low Tech und High Tech in der Realisierung übrig bleibt? Hoffentlich viel!

Unter dem Dach liegt dann die eigentliche BMW Welt, also Raum für eine permanente Performance der Produkte mit der „BMW Premiere", wie die Übergabezone der Fahrzeuge an die Abholer heißt. Aber auch vieles mehr. Coop Himmelb(l)au gliedern Erlebniswelten und -landschaften geschickt, lassen Raumzusammenhänge fließen und überraschen damit, wie der Grundriss innen zu einem immer wieder sich verändernden Raum wird, wie die BMW-Bühne zur vollen Wirkung kommt.

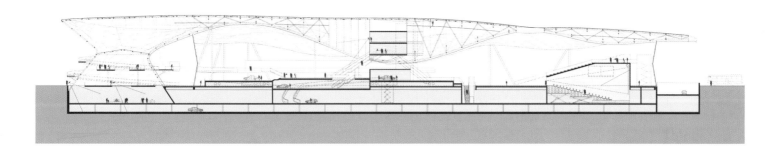

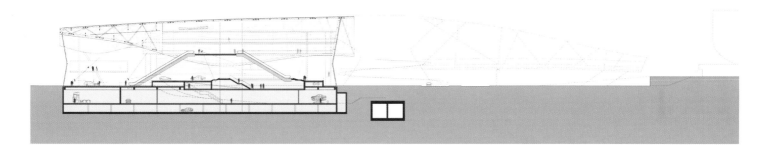

In the words of Coop Himmelb(l)au, this building will be a roof – without "façades as such, and as such it will present a sculptural entity" – and they will be relinquishing the age-old beam and column theme and charting new terrain.

Wolf Dieter Prix, the spokesman of Coop Himmelb(l)au, has his own flowery yet pithy formulation to describe this: he calls their intelligent roof system "clean energy cloud". When it is complete, this double-shell system will control ventilation, cooling and energy-generation. Even a solar power station could be integrated. It remains to be seen what will remain of this fantastic combination of low and high tech when it is completed. Let us hope it will be a lot.

Beneath this roof we then find the BMW world proper, with space for a permanent performance by the products and the "BMW Premiere", as the zone in which new vehicles are handed over to their owners is called. But that is not all. Coop Himmelb(l)au cleverly structure experiences and themescapes, and allow the relationship between spaces to flow and surprise. The result is an ever-changing room on the inside, and one in which the BMW stage unfolds to full effect.

Renderings und Schnitte für die BMW Welt, die ein lebendig gestaltetes, schwingendes Dach für die verschiedenen Facetten des Gebäudes bildet.
Renderings and sections for BMW Welt, which forms a vibrantly designed, dynamic roof for the various facets of the building.

Welche Autos brauchen wir, wie müssen sie sich in Zukunft noch verändern?

Wenn man das Auto z.B. in Los Angeles sieht, behaupte ich, dass es Teil des öffentlichen, urbanen Raumes geworden ist, weil man durch den Einsatz der Medien, wie Autoradio oder Autotelefon, jederzeit in Verbindung mit anderen Menschen ist. Wir müssen die Rolle des Autos neu interpretieren, andere Begriffe einführen, das Auto z.B. als öffentliche Kommunikationszelle betrachten, die Erotik und Bequemlichkeit der Geschwindigkeit genießen. Und nicht nur das Technoide in den Vordergrund stellen. Und ich glaube, dass die Autohersteller nicht nur ökologisch umdenken sollen, sondern auch platzsparend denken müssen. Ich kann mir vorstellen, dass man Teile der Autokabine flexibel halten muss. Also, wenn man alleine fährt, setzt man einen kleineren Teil ein, als wenn man zu viert fährt.

Autobauer sehen heute sehr häufig gute Architektur im Präsentationsumfeld ihrer Produkte. Welche Architektur passt zu schnellen Autos?

Architektur kann zwar nicht illustrative Verkaufsargumente, jedoch Identifikation leisten, weil sie dreidimensionale Kultur darstellt. Kultur, aber nicht nur Unternehmenskultur, sondern Stadtkultur, und zivilisatorische Kultur und, wenn sie besonders hoch angesiedelt ist, sagt sie die Zukunft der Ästhetik voraus. Seit dem 11. September 2001, als Architektur als Symbol begriffen und angegriffen wurde, ist eindeutig geworden, wie Architektur wirklich Zeichen für unsere Zivilisation setzt. Bezogen auf die Architektur der BMW Welt kann ich sagen, dass über die Ästhetik der Architektur, also nicht in Form eines Autos, ein Beitrag zur Stadtkultur und -gestalt geleistet und damit zum echten

Identifikationspunkt wird. Wir müssen dem Auto eine neue Bedeutung als Kommunikationsplatz geben – als Teil des öffentlichen Raums. Wir gehen also davon aus, dass das Auto sich vom Inhalt her ändern wird und nicht mehr nur einfach Fortbewegungsmittel sein kann. Daher müssen wir eine Ästhetik in der Auto-Architektur schaffen, die diese Flexibilität andeutet und transportiert.

Architektur für BMW = fließende Form?

Ich verweise auf unser Statement von 1976, dass offene Architektur nicht nur offene Räume bedeutet, sondern offenes Denken und offene Systeme zulässt. Das setzt einen dynamischen Prozess in Gang und besitzt damit Aufforderungscharakter zur Veränderung. Was die BMW Welt betrifft, sind wir dabei, das uralte System von Stütze und Balken aufzulösen und damit optisch die Schwerkraft in Frage zu stellen. Das ist heute zwar noch ein architekturtheoretisches Thema, wird aber in Zukunft eine große praktische Rolle spielen. Das Synonym für BMW ist ja, dass sie neue Techniken einsetzen, um neue Erkenntnisse, nicht nur über das Auto, sondern auch über neue Kraftstofferzeugung, über Energieerzeugung und das ganze Leben herauszufinden.

Gute Architektur entsteht nur dann, wenn man über 100 Prozent Leistung gibt. 180 Prozent denken, weil man Jahrzehnte vorausdenken muss: Erfunden 2001, gebaut 2005, steht die BMW Welt dann noch 30 bis 50 Jahre. Man muss also für mindestens 30 Jahre planen. Wir kennen so etwas von anderen Projekten, die vom Denken zehn Jahre alt sind und immer noch frisch dastehen. Solche Prozesse zu kommunizieren ist Aufgabe der Architekten.

→ **interview** interview

Architekt Wolf Dieter Prix von Coop Himmelb(l)au, Wien (im Bild rechts neben Helmut Swiczinsky) über das neue Verhältnis von Auto und Stadt und die Rolle der Architektur für die Zukunft des Autos.
Architect Wolf Dieter Prix of Coop Himmelb(l)au, Vienna (right, with Helmut Swiczinsky) on the new relationship between car and city, and the role of architecture in the future of the car.

What cars do we need, how will they have to change in the future?

If we look at the car in Los Angeles, say, then I would maintain that it has become part of public, urban space, because the use of media such as the car radio or car phone allows us to be in contact with others all the time. We have to reinterpret the role of the car and introduce new concepts, such as regarding the car as a public communication cell, or enjoying the eroticism and convenience of speed. We shouldn't simply concentrate on technological features. And I believe that car manufacturers not only have to rethink ecologically, but also have to think in space-saving terms. I could imagine having to keep parts of the passenger compartment flexible. So, if I'm driving alone, I'll use a smaller section than if there are four of us in the car.

Nowadays, car manufacturers are frequently using good architecture in the environment in which they present their products. Which architecture goes with fast cars?

While architecture cannot provide visual sales arguments, it can provide identification, because it presents three-dimensional culture. By culture I don't mean corporate culture, but urban culture and the culture of civilization. If its quality is particularly high, it can show the way forward for aesthetics. Since 11 September 2001, when architecture was understood and attacked as a symbol, it has become clear that architecture really does set signs for our civilization. In terms of the architecture of BMW Welt, I would say the architectural aesthetic (which

does not take the form of a car) makes a contribution to urban culture and form, and thus becomes a true identification point. We have to give the car a new significance as a place of communication – as part of public space. What we are assuming, then, is that the car will change in terms of content, and can no longer simply be a means of locomotion. That is why we have to create an aesthetic in car architecture which alludes to and conveys this flexibility.

Architecture for BMW = flowing form?

In a statement we made in 1976, we said that open architecture is not just open space, but also allows open thinking and open systems. This sets a dynamic process in motion, and is in this way a call to change. As concerns BMW Welt, we are in the process of abandoning the age-old column and beam system and thus of optically challenging gravity. While this is still a topic of architectural theory today, it will play a great practical role in the future. The counterpart of this for BMW is that they use new technologies to find out not just about the car, but also about new ways of generating fuel and power, and about life itself.

Good architecture only comes about if you give more than 100 per cent. The thought that goes into it has to be 180 per cent, because you have to think decades ahead: invented in 2001, built in 2005, BMW Welt will then stand for a further 30 to 50 years. So you have to plan for at least 30 years. We can see this in other projects: the ideas behind them are 10 years old, but they are still as fresh as ever. The architect's job is to put these processes across.

gedanken zur zukunft der motortecture
die stadt des 21. jahrhunderts und das auto

thoughts on the future of motortecture
the city of the 21st century and the car

gedanken zur zukunft der motortecture
die stadt des 21. jahrhunderts und das auto

Die kursiv gedruckten Sätze sind Ausschnitte und Zitate
aus einem Gespräch mit dem Architekten Meinhard von Gerkan.
Zeichnungen und Modelle von Luchao Harbour City

Rem Koolhaas lässt sein Kultbuch *Delirious New York* nicht mit
Autos, Autofahrern oder etwa Flugzeugen ausklingen, sondern
mit russischen Architekturstudenten, die allein durch ihre Muskel-
kraft eine autarke künstliche Insel durch den Atlantik von der Alten
zur Neuen Welt in Bewegung setzen. Die Interpretation dieses so
ganz eigenen mobilen Polis-Modells, das nicht an Land, sondern im
Wasser zur Blüte kommt, konterkariert den von Rem Koolhaas
entdeckten Manhattanismus, eines der wenigen realisierten Stadtmodelle des 20. Jahrhunderts. Hier nun könnte er im Zusammenhang
einer immer noch intellektuellen Stadtflucht stehen, die so gänzlich in Opposition zu den Realitäten dieser Welt steht. Die Automobilität in ihrer höchsten Form, dann, wenn sie nicht dem Gelderwerb,
also dem Pendeln zur Arbeit oder zum Ort des Handels dient, soll
die Menschen doch von der Zivilisation befreien; sie möglichst weit
weg von Menschen und Massen, von Technik und eben Auto führen.

Die totale Automobilität führt ins Gegenteil und schließlich zum Ent-
schluss einer Emanzipation vom Auto. Sie ist damit begründet, dass
die Orte, an denen das Auto regiert, nicht human genug sind und
immer unmenschlicher werden. Es ist bereits in diesem Buch darauf
hingewiesen worden, dass es trotzdem für eine geraume Zeit unmöglich sein wird, auf das Auto als Automobilitätsfaktor zu verzichten.
Und auf das Gebilde Stadt, jenen Moloch, der auf friedliche, feind-
liche und unheimliche Weise die Menschen immer häufiger anlockt,
gerade hier ihr kurzes (Auto-) Leben zu verbringen – auf die Stadt
als Lebensraum wollen wir genauso wenig verzichten. Im Gegenteil.
Ein verhängnisvoller, nicht aufzuhaltender Prozess der vermeint-
lichen Existenzsicherung führt den Städten global immer mehr
Menschen zu. Und macht aus ihnen Stadtlandschaften, die diese

thoughts on the future of motortecture
the city of the 21st century and the car

The sentences in italics are excerpts and quotations
from an interview with the architect Meinhard von Gerkan.
Drawings and models of Luchao Harbour City

Rem Koolhaas has his cult book *Delirious New York* end not with cars,
drivers or indeed aeroplanes, but with Russian architecture students
who, by muscle power alone, set a self-contained artificial island in
motion across the Atlantic, from the Old World to the New World.
The interpretation of this quite unique mobile model of a polis, which
flourishes on the water rather than on land, is a counterpoint to the
"Manhattanism" discovered by Rem Koolhaas, one of the few 20th
century urban models to have been put into practice. Here, it could be
seen in the context of intellectuals' flight from the city, which is so
completely opposed to the realities of our world. After all, in its highest
form, when it is not serving the purpose of earning money (i.e. commu-
ting to work or to the centres of commerce), automobility should liberate
man from civilization; it should lead him as far as possible away from
the masses, technology and, when all is said and done, the car.

Total automobility is self-contradictory, leading in the end to the decision
to free oneself from the car. This is because the places where the car
rules are not human enough and are becoming ever more inhuman.
Nevertheless, it has already been pointed out in this book that it will be
impossible for quite some time to do without the car as a means of
automobility. And we are just as unwilling to do without the city as our
habitat, the Moloch that exerts ever more frequently an attraction on
people – be it a peaceful, hostile or uncanny way – to spend their short
(automotive) lives there. On the contrary. In a fateful, inexorable process
that is supposed to safeguard their survival, more and more people are
flocking to the cities worldwide. That process is turning cities into cityscapes that are unworthy of the name, transforming metropolises into
ungovernable megalopolises. Even so, architects and town planners still
nurture the fond hope that they can do something about it. Perhaps they

Bezeichnung gar nicht verdienen, macht Metropolen zu einer unregierbaren Megalopolis.

Und dennoch bleibt und keimt bei Architekten und Stadtplanern die zarte Hoffnung, dass da was zu machen ist. Sein könnte. Das Auto hat sich geändert, ist vernünftiger, ja sogar ökologischer geworden. Was ist mit den Städten? Die alte Stadt ist perdu – sie war für die Ochsenkarren und den Fußgänger gestrickt, und so sollte es bleiben. Die neue hat versagt, auch dort, wo sie ganz neu geplant werden durfte. Es hat nur eine Hand voll Stadtneugründungen in der „frühen Autozeit", also zu Beginn und Mitte des 20. Jahrhunderts, gegeben. Solche Städte, die ganz Stadt mit allen ihren sozialen und kulturellen Aufgaben sind und nicht nur Stadtteile oder -erweiterungen. Sie waren wie das indische Chandigarh von Le Corbusier oder Brasilia (Oscar Niemeyer u.a.) von ideologischem und politischem Agitprop geprägt, von den besessenen Planern von Beginn an idealisiert, wie die entsprechenden Utopien und Denkmodelle, die sie vorher zu Papier gebracht hatten. Vielleicht sind die Architekten dann an diesem zu hohen Anspruch gescheitert oder es wirken die messbaren Ergebnisse im Vergleich zu den hohen Erwartungshaltungen kümmerlich.

Das ist Spekulation, belegbar hingegen ist, dass die Städtebauer das Auto unterschätzt hatten. Das betraf seine radikal anwachsende Zahl, die technischen Möglichkeiten sowie Beschleunigungsfähigkeit und Schnelligkeit. Alle geplanten Fahrspuren waren nach wenigen Jahren unzureichend, Straßenquerschnitte, Brücken- und Tunnelkapazitäten obsolet und der Parkraum zerstörerisch knapp, die Autos suchten sich ihren Platz irgendwo. Wir reden von der

can. The car has changed, has become more reasonable and, indeed, more ecological. What about our cities? The old town has been lost, it was designed for ox-carts and pedestrians, and it should stay like that. The new city has failed, even where it was possible to plan it from scratch. There were only a handful of newly founded cities in the "early car age", i.e. at the beginning and middle of the 20th century. By this I mean cities that were not just urban districts or extensions of cities, but true cities with all the social and cultural tasks this involves. Like Le Corbusier's Chandigarh in India or Brasilia (Oscar Niemeyer i.a.) they were marked by ideological and political agitprop, idealized right from the start by their obsessive planners, just like the utopias and theoretical models that they had committed to paper before building work. It may be that the architects were unable to live up to these ambitious aims, or the quantifiable results are pathetic by comparison with the great expectations placed in them.

We can only speculate about that. What is evident, however, is that urban planners underestimated the car. This is true of the tremendous increase in the number of cars and of cars' technical potential such as acceleration and speed. All the roads that were planned were inadequate within a few years, road sections and the capacities of bridges and tunnels were obsolete, and parking space was ominously scarce. We are talking about the early days of the automotive age. Things are different today. The peak has been passed or, in the economies still to reach their peak, it can be calculated and forecast better. And history wants new cities to be built and designed again.

There is a score to be settled, therefore. The Hamburg-based architect and town planner Meinhard von Gerkan is one of the three or four people

Frühzeit des Autozeitalters. Heute ist das anders, Höhepunkte sind bereits erreicht oder in den Volkswirtschaften, wo sie noch bevorstehen, besser kalkulierbar und prognostizierbar. Und die Geschichte will es, dass neue Städte wieder gebaut und ausgedacht werden.

Da ist also noch eine Rechnung offen. Der Hamburger Architekt und Stadtplaner Meinhard von Gerkan gehört zu den drei, vier Auserwählten in diesen Jahrzehnten, denen es vergönnt ist, eine ganze Stadt neu zu planen, eine Stadt größer als Braunschweig mit etwa 300.000 Einwohnern. Er tut es jetzt – nach all den Erfahrungen, die die Architektengenerationen vor ihm mit dem Auto in der Stadt gemacht haben.

gmp, von Gerkan, Marg und Partner, haben einen internationalen Wettbewerb gewonnen, die neue Stadt Luchao Harbour City zu entwerfen. Eine neue Hafenstadt zwischen Shanghai und einem gigantischen neuen Seehafen, der sozusagen mitten im Meer liegt. Es war eine Planung ohne Voraussetzung, am Meer, im Wasser, mit Dämmen oder vielen kleinen Kanälen, die ein unbesiedeltes Land durchziehen. *Nahezu jungfräuliche Bedingungen, die den Planer erfreuen und herausfordern. Man muss die Anknüpfungs- und Kristallisationspunkte aus sich selbst schöpfen,* wie Meinhard von Gerkan sagt. Zwei Dinge sind in Südchina, wo es eine Vielzahl von häufig unbekannten Millionenstädten gibt, die keinen Anfang und kein Ende besitzen, bedeutsam. *Diese zwei Dinge sind eine unverwechselbare Identität und städtische Bedingungen, die den Begriff der Urbanität verdienen und ein städtisches, verknüpftes Leben ermöglichen – Arbeiten, Einkaufen, Wohnen!*

in the last few decades to have been given the chance to plan a whole new city, a city bigger than Coventry or Nantes, with roughly 300,000 inhabitants. Now he is doing it – after all the experience of the city/car relationship gathered by the generations of architects before him. gmp (von Gerkan, Marg und Partner) won the international competition to design the new city Luchao Harbour City. This is to be a new city between Shanghai and a gigantic new seaport, floating in the middle of the sea, as it were. There were no conditions for the planning work; the city could be on the coast, in the water, with dykes or many small canals criss-crossing unsettled land. *Almost virgin conditions such as these are a joy and a challenge for planners.* As von Gerkan says, the planner has to use his own imagination to create urban contexts and cohesion.

In southern China, where there are many frequently obscure cities of a million and more inhabitants, with no definite beginning or end, two things are important. *These two things are an unmistakable identity and urban conditions that are worthy of urbanity and allow urban, cohesive existence – working, shopping, living.* Meinhard von Gerkan has thus also combined the opportunity of a completely new concept of a town with the car, because many people and a wide variety of uses will generate road traffic. As China continues its impetuous development, the individually used car will of course play just as great a role as in the West.

When all is said and done, the model proposed by von Gerkan is a "European city" in terms of its inspiration and qualities. It rejects the land-devouring American sprawl. Of course, Meinhard von Gerkan has been conditioned by his European and Hanseatic background, and by the way he approaches mathematical-geometric motifs and attitudes, an approach which has been honed and perfected over the past decades.

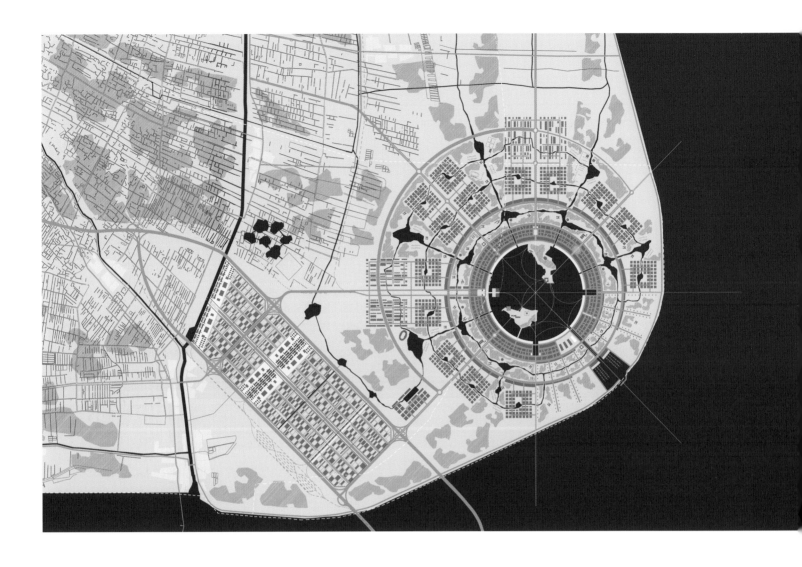

The result, for example, is that gmp came up with a far denser city than their international rivals dared to suggest, and than would have been necessary for the tract of land earmarked for the competition. It was a risk, but was in the end what gave them the edge. *Hamburg's topography is similar: With its Alster lake, Hamburg has a stretch of water at its centre, and everyone living there appreciates the qualities this bestows. We were forced to realize that a city on the coast, or rather on a cape, where the coastline changes direction, is a seaside city, but one whose ten-metre high dyke protecting it against typhoons and high tides prevents any visual contact between the city and the sea, unless you are on the top of the dyke or on the fourth floor of a building. There will be no view of the sea from its roads or squares.*

Because the location on the water's edge can only be used to a limited extent – an effect like that in Rio de Janeiro will not be possible – the architect generated a feeling of the city's water-borne identity by creating an artificial inland lake: *We propose a central lake, around which the central business area will be built. The form of the lake will be a precise circle. As we are dealing with an artificial feature, there is no need to clone natural forms. Instead we will use the metaphor of a drop falling into the water and creating concentric ripples. The theoretical or philosophical justification for their proposal is based on this metaphor and on an ordering principle whose point of reference is a compass or analogue clock face, which has quarter and eighth segments and a circular network of trunk roads fed by vertical minor roads. All of this refers to one common point of departure and orientation, the centre of the lake.*

Using a circular pattern such as this is both open to attack and inspired, but in the end the authorities approved and paid tribute to it, awarding

Damit hat Meinhard von Gerkan die Chance einer völlig neu kon-
zipierten Stadt auch mit dem Auto verknüpft, denn viele Menschen in
unterschiedlichen Nutzungen werden Straßenverkehr erzeugen. Das
individuell benutzte Auto spielt natürlich im sich ungestüm entwi-
ckelnden China eine ebenso große Rolle wie in westlichen
Ländern.

Es ist nach wie vor ein Modell, das aufgrund seiner Vorbilder und
Qualitäten „europäische Stadt" genannt wird, das von Gerkan vor-
schlägt. Es sind nicht die landgreifenden amerikanischen Unstädte
erwünscht. Meinhard von Gerkan wird natürlich durch seinen
europäischen und hanseatischen Hintergrund geprägt. Und durch
einen im Lauf der Jahrzehnte verfeinerten und präzisierten Umgang
mit mathematisch-geometrischen Motiven und Haltungen. Das
hat dazu geführt, dass gmp beispielsweise weitaus stärker die Stadt
verdichtet als die internationale Konkurrenz es sich traut und das
weitläufige Wettbewerbsgebiet es erzwingen würde. Das war ein
Risiko, hat sich aber schließlich als Matchwinner erwiesen. Das
Gleiche gilt auch für die Hamburger Topografie: *Hamburg hat mit
der Außenalster einen See mitten im Zentrum, und jeder dort weiß
dessen Qualitäten zu schätzen. Wir mussten erkennen, dass eine
Stadt am Meer, also in einer Kapsituation, dort, wo die Küstenlinie
ihre Richtung ändert, zwar eine Stadt am Meer ist, aber ein zehn
Meter hoher Damm als Schutz gegen Taifune und Hochwasserfluten
jeden visuellen Kontakt vom Stadtraum zum Meer unterbindet,
es sei denn, man befindet sich auf der Deichkrone oder im vierten
Obergeschoss. Von den Straßen und Plätzen her wird man das
Meer nicht sehen können.*

it first prize. Of course, everyone knows that there is a risk that a circular
pattern will cause some boredom. At the same time, however, it makes
it easy for people to get their bearings. They do so in a subliminal way,
out of a collective consciousness. This also holds true for road traffic.
*Definitely. We can see that the logic of geometry solves many things
in a self-evident, meaningful way. Take the public transport system, for
example. If the town has been given a circular structure, a commuter
railway can be designed to follow a circular route, as can a tram or
underground line. In our case, this is a tram which simply drives along
this circle line. One clockwise, the other anti-clockwise. Anyone using
this tram can in effect get on a tram in either direction and still reach his
destination. In other words, this way of organizing traffic is extremely
self-evident and very logical.*

The rest of the way the system is ordered is simple, as the city is
arranged like a clock face, with stops at twelve, one, two, three and so
on. Radial roads lead out from this ring, providing access to the residential districts. Islands 640 metres square lie between these roads, each
island with about 13,000 inhabitants. The quarters formed in this way
make up independent, small communities. Building on a scale like this
is human, and the traffic situation that results from them is manageable,
possibly an alternative to those earlier city models which proved to
be badly designed when they were actually built. *I first visited Brasilia
roughly 15 years ago, a long time after it was built. I stayed there for
several days, in a mood which was a mixture of interest and misgiving.
What I learned there was that this town-planning approach was an
absolute cul de sac, and that Brasilia is so utterly a city for the car that
people can no longer move in this city. First, it is practically impossible
to cross the large arterial roads, and second, none of the things you need*

Weil die Lage am Wasser nur eingeschränkt nutzbar ist, also kein Effekt wie in Rio de Janeiro entsteht, hat der Architekt die Identitätsstiftung durch Wasser selbst durch einen künstlichen Binnensee generiert: *Wir schlagen einen zentralen See vor, darum wird sich ringförmig der Central-Business-Bereich legen. Die Form des Sees folgt exakt der Kreisform, wir haben es ja mit künstlichen Setzungen zu tun, brauchen keine natürlichen Formen zu klonen, sondern nutzen die Metapher eines Tropfens, der ins Wasser fällt und konzentrische Kreise zieht. Die theoretische oder philosophische Begründung fußt auf dieser Metapher und auf einem Ordnungsprinzip, das sich auf einen Kompass oder eine analoge Uhr bezieht, Viertel- und Achtelsegmente besitzt und kreisförmig geführte übergeordnete Straßen und dazugehörige, senkrecht verlaufende Erschließungsstraßen hat. Und alles bezieht sich wieder auf den gleichen Ausgangs- und Orientierungspunkt, eben den Mittelpunkt des Sees.*

Diese Setzung im Zeichen des Kreises ist gleichermaßen angreifbar wie genial, die chinesischen Behörden haben sie schließlich durch die Vergabe des ersten Preises sanktioniert und gewürdigt. Jeder weiß natürlich, dass eine kreisförmige Organisation Gefahr läuft, gewisse Langeweile zu verbreiten, aber eine grundsätzliche Sicherheit in der Orientierung leistet. Und zwar eine unterschwellige, eine aus dem kollektiven Gedächtnis heraus. Das gilt auch für den Straßenverkehr. *In jeder Hinsicht. Es zeigt sich, dass die Logik der Geometrie viele Dinge auf eine selbstverständliche und sinnfällige Weise löst. So auch das System des öffentlichen Nahverkehrs. Wenn also die Stadtstruktur kreisförmig angelegt worden ist, kann eine Regionalbahn in einem Ring angeordnet werden, ebenso*

for everyday life can be reached on foot. The city is so dominated by flowing and stationary traffic that urbanity has withdrawn into a few pockets. The most urbane place in Brasilia is its main bus station, under an expressway.

This was a crucial experience, and von Gerkan used the lessons he learned from it for his new "ideal city". While he does not want Luchao Harbour City to bar traffic, traffic should not dominate the city either. This is almost like squaring the circle. There are two radial connections to the regional motorway system. Two major ring roads, with no buildings along them, accommodate long-distance traffic. The innermost ring road, around the inland lake, is access only – there is no through traffic. The adjoining business district is only accessible for those working there, for deliveries and emergency vehicles, and can otherwise be traffic-calmed. All the access roads are for traffic originating there or with business there, types of traffic whose scale is human and which are forced to mix with pedestrian traffic.

All of this is still an ideal in the planning stage, but the general idea is clear: *the new city has the characteristics of a European city, with clearly defined roads, and strategically placed squares and spaces with public buildings. And each of the clearly defined sub-communities is based on a 640 metre square plot, again with a small lake at its centre, resulting from the existing system of waterways. Each of the communities has its own definite centre, with schools, shops and other infrastructure.*

So much for the theory. It remains to be seen whether a clearly defined, dense structure of blocks will be accepted in China, where it is regarded as "impossible" to live in houses built in an East-West direction. Taken

eine Straßenbahn oder eine U-Bahn. In diesem Fall ist es eine Straßenbahn, die einfach diesen Ring entlangfährt. Einmal im Uhrzeigersinn und einmal gegen den Uhrzeigersinn. Wenn man diese Bahn benutzt, kann man eigentlich in beide Richtungen einsteigen und kommt trotzdem immer am Ziel an. Also ist es eine sehr selbstverständliche, sehr logische Organisation des Verkehrs.

Das weitere Ordnungssystem ist einfach, die Stadt ist wie eine Uhr organisiert, mit Haltestellen bei zwölf, eins, zwei, drei und so weiter. Und von diesem Ring ausgehend führen dann radial Straßen der Feinerschließung in die Wohnquartiere. Dazwischen liegen Inseln mit 640 mal 640 Metern, jeweils mit etwa 13.000 Einwohnern in einem solchen Quartier, das jeweils eine selbstständige Kommune oder kleine Gemeinde bildet. Ein solcher Maßstab ist human, führt zu überschaubaren Verkehrssituationen und bildet möglicherweise die Alternative zu früheren Stadtmodellen, die sich bei der Realisierung als Fehlkonstruktionen erwiesen hatten. *Ich habe erst lange Zeit, nachdem Brasilia gebaut worden ist, vor etwa 15 Jahren, diese Stadt besucht. Mit einem gewissen Interesse, aber auch mit einer gewissen Reserviertheit habe ich mich dann mehrere Tage dort aufgehalten. Und gelernt, dass dieser Weg der Stadtplanung ein absoluter Irrweg und Brasilia in einem solchen Maße eine Stadt für das Auto ist, dass sich der Mensch in dieser Stadt gar nicht mehr bewegen kann. Erstens kommt man über die großen Verbindungsstraßen gar nicht hinweg, zweitens kann man nichts fußläufig erreichen, was man für den alltäglichen Bedarf braucht. Die Stadt ist vom fließenden und ruhenden Verkehr so dominiert, dass Urbanität sich auf Restflächen zurückzieht. Der urbanste Ort Brasilias ist eine zentrale Busstation unterhalb einer Schnellstraße.*

Von Gerkan hat aus diesem Schlüsselerlebnis für seine neue „Idealstadt" gelernt. Luchao Harbour City soll den Verkehr nicht aussperren, er soll jedoch die Stadt nicht dominieren. Das ist so etwas wie die Quadratur des Kreises: Es gibt zwei radiale Anbindungen an das überörtliche Autobahnsystem. Zwei übergeordnete, bebauungsfreie Ringstraßen fangen den Fernverkehr auf. Der innerste Ring am Binnensee dient nur dem Anliegerverkehr – ohne Durchgangsverkehr. Der anschließende Geschäftsdistrikt wird nur für Anlieger-, Liefer- und Notverkehre erreichbar und ansonsten ein verkehrsberuhigter Distrikt sein können. Alle Erschließungsstraßen dienen dem Ziel- und Quellverkehr, Verkehrsarten, die human bleiben und sich zwangsläufig mit Fußgängerverkehren mischen.

Noch ist dies alles idealistischer Planungszustand, aber die Dinge liegen klar: *Die neue Stadt trägt die Merkmale der europäischen Stadt mit klaren Straßenprofilen, mit pointiert gesetzten Platzräumen mit öffentlichen Gebäuden. Und die klar umgrenzten Teilkommunen sind immer auf ein Feld von 640 mal 640 Metern bezogen, besitzen in ihrem Zentrum wieder einen kleinen See, der sich aus dem Geflecht der vorhandenen Gewässer ergibt. Die Kommunen besitzen einen klaren Ortskern mit Schulen, Läden und anderer Infrastruktur.*

So weit die Theorie. Ob eine klare, dichte Blockstruktur sich in China durchsetzen lässt, wo es sozusagen als unmöglich gilt, in Häusern zu wohnen, die nach Osten oder Westen liegen, ist abzuwarten. Von Gerkan setzt, wenn man es zusammenfasst, auf eine vorzunehmende Durchmischung der entmischten Stadt, soll heißen, der Verkehr wird abnehmen, weil immer mehr Menschen auch dort wohnen können, wo sie arbeiten. Natürlich wird das stark

emittierende Gewerbe an die Ränder verbannt, aber selbst in China gibt es in einer neu zu bauenden Stadt davon nicht mehr so viel.

Auch wird es Stadtgebiete geben, die nur für das Wohnen reserviert sind, aber je mehr man sich dem Zentrum nähert, umso mehr wird Wohnen und Arbeiten in einer recht hohen Verdichtung gemischt: Dies ist sicher im Sinne vor allem der neuen niederländischen Urbanisten, die eine weitere Zersiedlung und den Landschaftsfraß stoppen wollen. Dies ist nicht das Modell der amerikanischen Stadt! *Wo die Verdichtung am größten ist, wird das Wohnen nicht über-wiegen, aber noch 30 bis 40 Prozent ausmachen, mit Wohnungen für Singles, die es natürlich auch in China gibt.*

Auch wenn China unter dem Verstädterungsprozess leidet, soll heißen, alles in die Städte strebt, existiert dort inzwischen schon ein Konkurrenzkampf um die wohlhabenden und jungen Bürger. Luchao Harbour City hat ein entsprechend attraktives Konzept, das ein hoch entwickeltes Stadtgefüge in eine intakte, erlebbare Land-schaft setzt und, so ist zu hoffen, das Auto zügelt, aber zulässt. Ob es klappt – schon in wenigen Jahren werden wir es wissen, weil China dieses Projekt mit noch nie gesehener Geschwindigkeit ver-wirklichen wird. Und man wird sehen, ob die übrige Welt von China lernen kann, das für seine Weltausstellung in Shanghai 2010 das Motto ausgegeben hat: „Better City, better Living"! Rem Koolhaas übrigens glaubt, dass die wesentlichen stadtentwicklungspoliti-schen Meilensteine in Asien und vor allem in China gesetzt werden – als vernünftige Utopien.

as a whole, von Gerkan hopes that his discretely organized city will change to one of mixed use; in other words, that traffic will decrease because more and more people are able to live where they work. Of course, extremely polluting industry will be pushed out to the margins, but even in China there is no longer so much industry of this kind in a new city.

There will also be districts that are reserved solely for residential pur-poses, but the closer one gets to the centre, the more living and working will be mixed, at a quite high level of density. This is without doubt something that would find favour with the new Dutch urbanists, who want to stop more sprawls and more countryside being eaten up. It is certainly not the model of an American city. *Residential use will not predominate where density is highest, but will still account for between 30 and 40 per cent of use, with apartments for singletons, who you will of course also find in China.*

Even if China is feeling the burden of rural exodus, with everyone on the move to the cities, those cities are meanwhile also competing with each other for well-off, young citizens. Luchao Harbour City has the attractive concept that this requires, setting a highly developed urban structure in an intact, enjoyable landscape and – or so we hope – curbing the car while tolerating it. Will it work? In a few years at the latest we will know the answer, because China will put this project into practice with unpre-cedented speed. And we will see whether the rest of the world can learn from China, which has chosen the motto "Better City, Better Living" for the world exhibition in Shanghai in 2010. Rem Koolhaas, for his part, holds that the major urban planning milestones of the future will be set in Asia, and above all in China – as rational utopias.

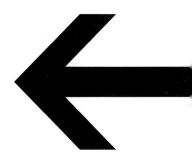

→ **bibliografie** bibliography

Volker Albus, Mateo Kries,
Alexander von Vegesack (Hrsg.)
Automobility.
Was uns bewegt
Weil am Rhein 1999

Jonathan Bell
Carchitecture.
When the car and the city collide
Basel, Boston, Berlin 2001

Gernot Brauer (Hrsg.)
Architektur als Markenkommunikation.
Dynaform + Cube
Basel, Boston, Berlin 2002

Peter Cachola Schmal (Hrsg.)
Digital real: Blobmeister.
Erste gebaute Projekte
Basel, Boston, Berlin 2001

Die neuen Architekturführer.
Jahrgang 2000/2001.
11 neue Architekturführer über 11 neue Bauten
Berlin 2001

Kristin Feireiss (Hrsg.)
Ingenhoven Overdiek und Partner –
energies
Basel, Boston, Berlin 2002

Kristin Feireiss, Hans-Jürgen Commerell
BMW Erlebnis- und Auslieferungszentrum.
Realisierungswettbewerb
o. O. 2002

Meinhard von Gerkan
von Gerkan, Marg und Partner –
Architecture 1997–1999
Basel, Boston, Berlin 2000

Meinhard von Gerkan
von Gerkan, Marg und Partner –
Architecture 1999–2000
Basel, Boston, Berlin 2002

Gunter Henn (Hrsg.)
Audi-Forum Ingolstadt.
Tradition und Vision
München, London, New York 2001

Gunter Henn
Form follows Flow.
Modulare Fabrik Skoda
München 1998

Ingenhoven Overdiek und Partner,
KMS Team (Hrsg.)
Architektur und Design.
Neue Synergien
Basel, Boston, Berlin 2001

Rem Koolhaas
Delirious New York.
Ein retroaktives Manifest für Manhattan
Aachen 1999

Richard C. Levene,
Fernando Márquez Cecilia (Hrsg.)
Neutelings Riedijk
Madrid 1999

Dirk Meyhöfer, Ullrich Schwarz (Hrsg.)
Architektur in Hamburg.
Jahrbuch 2002
Hamburg 2002

Renzo Piano
Mein Architektur-Logbuch
Ostfildern-Ruit, 1997

Bernd Polster
Tankstellen.
Die Benzingeschichte
Berlin 1982

Rainer Stommer (Hrsg.)
Reichsautobahn.
Pyramiden des Dritten Reichs.
Analysen zur Ästhetik
eines unbewältigten Mythos
Marburg 1982

Wolfgang Weyrauch (Hrsg.)
Alle diese Straßen
München 1965

cover photo:
Karsten de Riese, Bairawies

motortecture
ein jahrhundert architektur und automobilität
motortecture
one century of architecture and automobility

7: Cornelius Scriba, Hamburg
9 top: Photo Deutsches Museum München
9 below, 10, 13, 25 top: DaimlerChrysler Konzernarchiv
11: SVT-Bild/DAS FOTOARCHIV
12: Esso A.G.
14 top: SVT-Bild/DAS FOTOARCHIV
14 below: © artur/Michel Denancé/Archipress
15: Thomas Mayer/DAS FOTOARCHIV
17: © Architectural Association,
Photographer: Yerbury, F.R.
18: Michael Nath Fotodesign
19, 20, 21: © FLC/VG Bild-Kunst, Bonn 2003
23: DAS FOTOARCHIV
24: Jochen Tack/DAS FOTOARCHIV
26: Reimer Wulf
27 top: Deutsches Architektur Museum,
Frankfurt am Main
27 below: Jörg Hempel
28: MVRDV

an der straße, für die straße:
brücke, tankstelle, bahnhof
along the road, for the road:
bridge, filling station, tram station

31: Tony Garbasso, Andrea Martiradonna
32: © artur/Sylvie Bersout/Archipress
33: © artur/Michel Denancé/Archipress
34, 35: © artur/Tomas Riehle
36–39: von Gerkan, Marg und Partner
40 top: Bänziger + Bacchetta + Partner, Zürich
40 below, 41: Christian Menn
42, 43: © artur/Roland Halbe
44, 45: Christian Richters
46–49: Tony Garbasso, Andrea Martiradonna
50, 51: Georges Fessy, Paris
52–55: Ingo Scheffler, Berlin
56, 57: © artur/Roland Halbe
58: Schlaich, Bergermann und Partner

häuser für autos:
parkhaus, feuerwehr
buildings for cars:
garage, fire station

59–65: Kozo Takayama
66, 67: © artur/Wolfram Janzer
68–77: Christian Richters
78: Petry + Wittfoht

stadt für autos ohne autos:
die autostadt wolfsburg
motown without cars:
autostadt in wolfsburg

79: Karsten de Riese, Bairawies
80, 82, 83, 85 top, 86, 87 left, 89, 91 top, 92,
94–96, 97 left, 98, 100–102, 103 left:
Werner Huthmacher, Berlin
81, 88 left: Stefan Müller-Naumann, München
84, 103 right: Fetzi Baur, Filderstadt, www.fetzibaur.com
85 below, 88 right: Ulrich Schwarz, Berlin
87 right, 90, 91 below, 93 below, 97 right:
Dieter Leistner, Mainz
93 top, 99: Autostadt GmbH, Wolfsburg
104, 105: Doug Snower
106: © Wilfried Dechau, Stuttgart
107, 108: Henn Architekten, München

wo die autos herkommen:
fabrik und manufaktur
where the cars come from:
plants and manufactory

109, 113–117: Stefan Müller-Naumann, München
110, 111: H. Leiska
118, 119 below, 120 sections, 123 top:
Renzo Piano Building Workshop
119 top, 120–122: Architekturfoto Peter C. Horn
123 below: DaimlerChrysler AG
124–127, 128 top, 129 top, 132:
Werner Huthmacher, Berlin
128 below, 133: Karsten de Riese, Bairawies
129 below, 130, 131: H.G. Esch, Hennef/Sieg
134: Konrad R. Müller, Königswinter

corporate architecture:
showroom und messepavillon
corporate architecture:
showrooms and fair stands

135, 136, 138, 139: H.G. Esch, Hennef/Sieg
137: Andreas Keller, Kirchentellinsfurt
140–143: Saab Deutschland GmbH, Bad Homburg
144–149: Mercedes-Benz Niederlassung Berlin
150 top, 151: ABB Architekten/Bernhard Franken
150 below: Fritz Busam, Berlin
152, 153 below, 154–159:
© friedrich busam/architekturphoto
153 top: Jörg Hempel
160: Franken Architekten, Frankfurt am Main

gemischte angebote:
museen und weitere projekte
mixed offerings:
museums and other projects

161, 168, 169 below, 170, 171, 176–179:
Stefan Müller-Naumann, München
163, 166, 167: H. Leiska
164–165: O. Heßner
169 top: Henn Architekten, München
172, 173 top, 175: UN Studio: Ben van Berkel,
Tobias Wallisser with Marco Hemmerling,
Andreas Bogenschütz, Alexander Jung, Björn Rimner,
Arjan van der Bliek, Götz Peter Feldmann,
Arjan Dingsté, Wouter de Jonge, Rombout Loman,
Anna Carlquist
173 below, 174: DaimlerChrysler AG
180–187: BMW AG
188: Aleksandtra Pawloff

gedanken zur zukunft der motortecture
die stadt des 21. jahrhunderts und das auto
thoughts on the future of motortecture
the city of the 21st century and the car

189: DAS FOTOARCHIV
193, 196, 197: von Gerkan, Marg und Partner

**Wir haben uns bemüht, alle Quellen und
Rechteinhaber zu ermitteln und vollständig
anzugeben. Sollten dennoch Angaben fehlen,
bitten wir um Benachrichtigung.**

We have made every effort to establish sources
and the holders of rights, and to give full
details. Should details nevertheless be missing,
please contact us.

→ **impressum** publishing details

Konzeption und Text
Dirk Meyhöfer, Hamburg

Übersetzung
Philip Mann

Redaktion
Anja Schrade

Gestaltung
Angela Kühn, Hamburg

Coverfoto
Karsten de Riese, Bairawies

Lithografie
ctrl-s prepress GmbH

Produktion
avcommunication
Gunther Heeb

Druck
Leibfarth & Schwarz,
Dettingen / Erms

Bibliografische Information der Deutschen Bibliothek:
Die Deutsche Bibliothek verzeichnet diese Publikation in der
Deutschen Nationalbibliografie; detaillierte bibliografische
Daten sind im Internet über http://ddb.de abrufbar.

© **Copyright 2003**
avedition GmbH, Ludwigsburg

© **Copyright für Fotos und Pläne**
bei den Unternehmen und Fotografen

Alle Rechte, insbesondere das Recht der Vervielfältigung,
Verbreitung und Übersetzung vorbehalten. Kein Teil des Werkes
darf in irgendeiner Form (durch Fotokopie, Mikrofilm oder ein
anderes Verfahren) ohne schriftliche Genehmigung reproduziert
oder unter Verwendung elektronischer Systeme verarbeitet,
vervielfältigt oder verbreitet werden.

ISBN 3-929638-77-0
Printed in Germany

avedition GmbH
Königsallee 57
71638 Ludwigsburg
www.avedition.de
kontakt@avedition.de

Original idea and text
Dirk Meyhöfer, Hamburg

Translation
Philip Mann

Editing
Anja Schrade

Design
Angela Kühn, Hamburg

Cover photo
Karsten de Riese, Bairawies

Lithography
ctrl-s prepress GmbH

Production
avcommunication
Gunther Heeb

Printed by
Leibfarth & Schwarz,
Dettingen / Erms

Bibliographic information published by Die Deutsche Bibliothek:
Die Deutsche Bibliothek lists this publication in the Deutsche
Nationalbibliografie; detailed bibliographic data are available
on the Internet at http://ddb.de.

© Copyright 2003
avedition GmbH, Ludwigsburg

© Copyright of photos and plans with
individual photographers and companies

All rights reserved, in particular the right of duplication, distribution
and translation. Without our prior permission in writing, no part
of this publication may be reproduced in any form (by photocopying,
on microfilm or using any other technique), or processed, copied
or distributed using electronic systems.

ISBN 3-929638-77-0
Printed in Germany

avedition GmbH
Königsallee 57
71638 Ludwigsburg
www.avedition.de
kontakt@avedition.de

→ **dank** acknowledgements

**Verlag und Autor danken den beteiligten Firmen,
Architekten, Agenturen und Fotografen für die zur
Verfügung gestellten Bilder und Materialien.**

**Besonderer Dank
für die freundliche Unterstützung gilt:**

**DaimlerChrysler AG / Stuttgart:
Peter Pfeiffer, Leiter Design Mercedes-Benz Design
Harald Leschke, Advanced Design Projekte / Corporate Design
Peter Hilken, Leiter Architekturcenter
Hans-Harald Hanson, Strategisches Design**

HENN Architekten / München

Saab Deutschland GmbH / Bad Homburg

Autostadt GmbH / Wolfsburg

The publisher and author wish to thank those companies,
architects, agencies and photographers who have provided
images and material.

We would like to extend our
particular thanks to:

DaimlerChrysler AG / Stuttgart:
Peter Pfeiffer, Head of Design at Mercedes-Benz Design
Harald Leschke, Advanced Design Projects / Corporate Design
Peter Hilken, Head of the Architecture Centre
Hans-Harald Hanson, Strategic Design

HENN Architekten / Munich

Saab Deutschland GmbH / Bad Homburg

Autostadt GmbH / Wolfsburg